Posterolateral Stabilomics Proceduralism for Thoracolumbal Pulverization Comminution

Kamiya Hiroshi

Posterolateral Stabilomics Proceduralism for Thoracolumbal Pulverization Comminution

First Edition December 2025

Written by Kamiya Hiroshi

CONTENTS

LIST OF FIGURES

LIST OF TABLES

INTRODUCTION

Fractures of the thoracic and lumbar spine account for nearly 90% of all spinal fractures. Thoracolumbar fractures usually result from high-energy trauma, most commonly falls from height or road traffic accidents. These injuries typically produce vertical compression of a slightly flexed spine, although rotational, shear, and extension forces may also lead to different fracture patterns. In addition to being a significant cause of morbidity and mortality, neurological impairment is associated with approximately 20% of these injuries.

Most patients with thoracolumbar fractures belong to the active and productive age group. Over the past few decades, considerable advances have been made in intraoperative monitoring, diagnostic imaging, and the development of more stable fixation devices. In addition, the use of steroids to reduce secondary injury to the central nervous system has received substantial attention.

Despite these advances, the management of thoracolumbar fractures continues to present challenges for orthopedic surgeons. Several surgical treatment options have been developed for thoracolumbar burst fractures, including short- or long-segment posterior pedicle screw fixation, direct anterior decompression through corpectomy, and combined anterior–posterior spinal approaches.

The choice of surgical techniques, however, is still up for debate. Regarding the use of short or long posterior fixation constructs, there is controversy over the best course of therapy for thoracolumbar burst fractures.[9,10] Long posterior fixation with pedicle screws and rods two levels above and below the fracture level improves fixation, but it also puts more strain on the lower discs and may result in unnecessary instrumentation. On the other hand, short posterior fixation limits the number of fused segments and prevents excessive loads on the nearby discs by employing pedicle screws with linked rods one level above and below the fracture level.[11,12] The use of short posterior fixation for burst fractures has been documented, but literature reviews on the procedure reveal significant rates of fixation failure and kyphotic collapse.[13]

There have been numerous trials comparing posterior short segment fixation to posterior long segment fixation in terms of clinical and radiological effectiveness.[14,15] With this background, our research aims to compare the short segment posterior stabilisation and long segment posterior stabilisation for the management of thoracolumbar burst fractures.

AIM OF THE STUDY

To study and analyse the Functional and Radiological Outcome of unstable Fractures of Thoracolumbar spine treated by short segment posterior stabilization and long segment stabilization.

REVIEW OF LITERATURE

The vertebral column comprises of discs and vertebra, supported by paraspinal muscles and ligaments. There are thirty-three vertebrae in toto which includes cervical – 7, thoracic – 12, lumbar - 5, sacral - 5 and coccygeal – 4. There are twenty-four mobile segments in a vertebra.[16] A neural arch and the body make up a vertebra. Neural arch gives rise to two pedicles and two laminae which come together to form the spinous process. The superior and inferior articular facet joints, along with two transverse processes laterally, allow the spine to move. Spinal cord is shielded by bony vertebral column. The intervertebral disc, which is located between the two vertebral bodies, is made up of an inner layer of nucleus pulposus and an outer layer of annulus fibrosus.[1, 16]

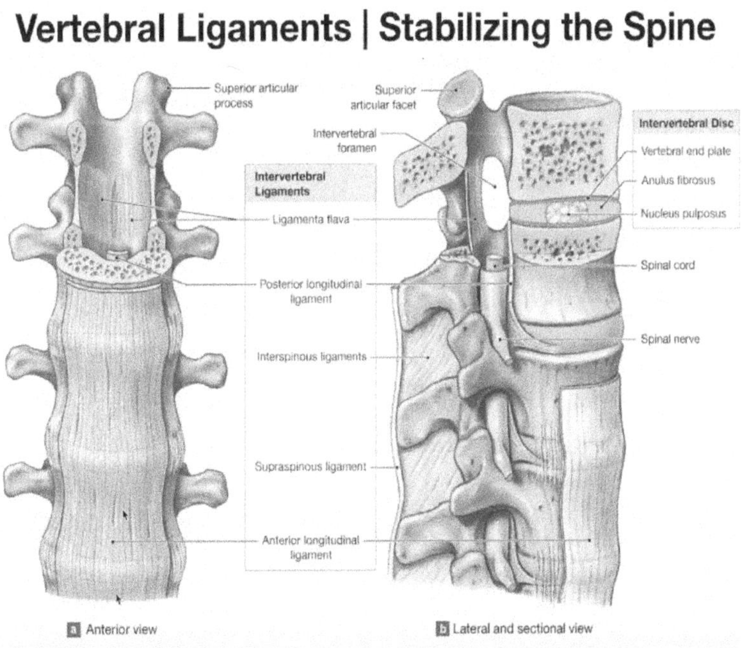

FIGURE 1: Vertebral ligaments

Anatomy of pedicle:

The strongest component of the vertebra is the pedicle, which joins the body and the posterior spinal segments. The pedicle is made of a strong shell of cortical bone and a core of cancellous bone. There are various sizes and angles for pedicles. The pedicle's width is often smaller than its height. The pedicle is widest in the thoracic region at D11, while the greatest pedicle in the lumbar region is at L5. Between D3 and L5, the pedicle's horizontal diameter (width) and vertical diameter (height) both expand from 7 to 15 and 16 millimetres, respectively. From D4 to L4, the pedicle runs virtually sagittally. At the thoracolumbar junction, the angulation is around 10 degrees, and at L5, it is about 30 degrees. On the medial side below L1, the cauda equina is extremely near to the pedicle.[17]

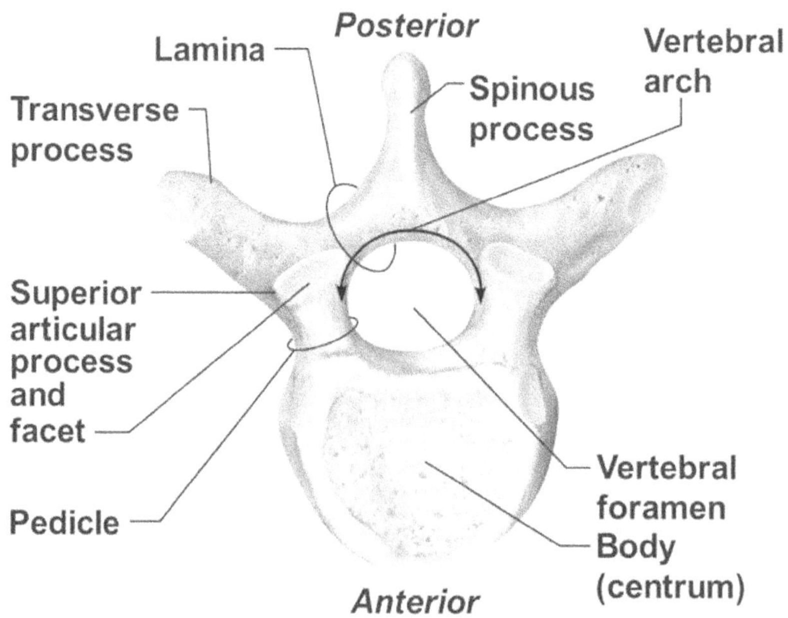

FIGURE 2: Vertebra anatomy

Blood supply of Spinal Column:

Blood supply to the thoracolumbar spine is from the posterior intercostal and lumbar arteries. Veins create a plexus through the vertebral column, both internal & exterior to the vertebral column, with no valves. There are anterior & posterior external vertebral veins. They connect to the internal plexus of veins by taking branches from the vertebral bodies. The internal plexus drains the vertebral bodies and spinal cord and has two anterior and two posterior branches. The basivertebral veins drain the posterior foramina of the vertebral bodies. The vertebral venous plexus is where the intervertebral veins, which predominantly drain the spinal cord, terminates.[18]

Lymphatic drainage:

The lymphatics from the thoracic region drain into the intercostal nodes, whereas lymphatics from the lumbar colon drain into the paraaortic and retroaortic nodes.[18]

Innervation:

Spinal nerves innervate the vertebral column. The sympathetic system innervates through grey rami communicantes. Facet joints & periosteum of the posterior bony elements are supplied by spinal nerve.[17,18]

Neuroanatomy of spinal cord:

The spinal cord extends from the foramen magnum to the lower border of L1. It has an oval shape. Duramater, arachnoid mater, and piamater surround it. The subarachnoid space includes cerebrospinal fluid and is located between the dura and the piamater. It ends in the conus medullaris, from wheree the filum terminale inclines. It is made up of the cortical white mater and grey mater in the inner aspect. The dorsal root is formed by roots arising from the posterolateral sulcus, which culminates in ganglion before it joins the ventral root to form spinal nerve. The ventral root is formed by roots arising from the anterolateral sulcus. Eight cervical, twelve thoracic, five lumbar, five sacral, and one coccygeal spinal nerve pairs are among the 31 pairings that make up the spine. Through their separate neural foramina, the spinal nerves below L1 exit as the Cauda equina.[19]

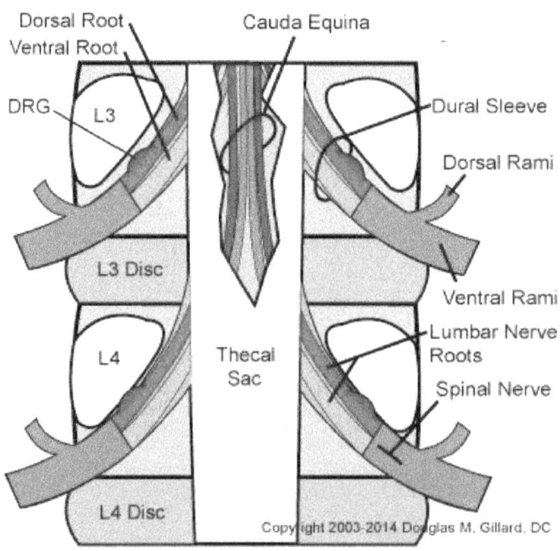

FIGURE 3: Cord and its roots

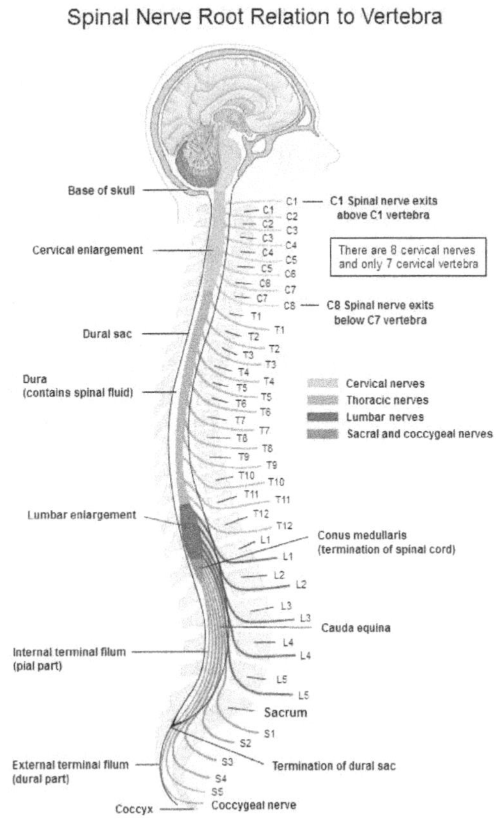

FIGURE 4: Level of vertebra and nerve roots

VASCULAR SUPPLY OF SPINAL CORD:

Segmental and longitudinal blood vessels both serve spinal cord. Longitudinal vessel is composed of an anterior spinal artery & 2 posterior spinal artery. Anterior spinal artery is derived from the vertebral artery. Ipsilateral vertebral artery or posterior inferior cerebellar artery supplies posterior spinal artery. Spinal cord is additionally supplied by the segmental medullary feeder arteries.[20] The anterior medullary feeder arteries range from 2 to 17 whereas the posterior medullary feeder arteries range from 6 to 25.

The largest anterior medullary feeder artery is the Adamkiewicz artery. Anterior spinal artery supplies blood to the anterior two thirds of the spinal cord. One branch from posterior spinal artery & pial plexus supplies the posterior 1/3rd of the cord. Each vertebral level's intraspinal and extraspinal structures are supplied by a pair of segmental arteries. The distribution points are produced by the numerous branching of the segmental arteries at the intervertebral foramen. Between T4 and T9, the thoracic cord's blood supply is minimal.[16, 20]

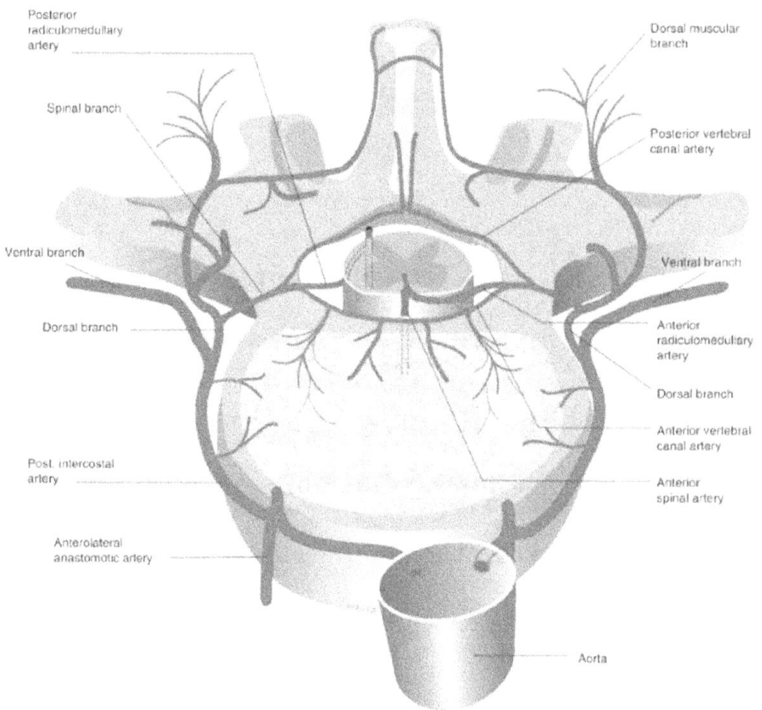

FIGURE 5: Arterial supply

The spinal cord's venous drainage varies widely. Veins in the spinal cord and those in the Batson plexiform network are two different types. The Batson plexus is where the spinal cord's veins enter. The extradural vertebral venous plexus, extravertebral venous plexus, and the veins of the bony bones of the spinal column make up the three components of the Batson plexus of veins. Metastatic spread from the pelvis to the spinal column is made possible by their direct communication with the venous system that drains the head, chest, and abdomen.[20]

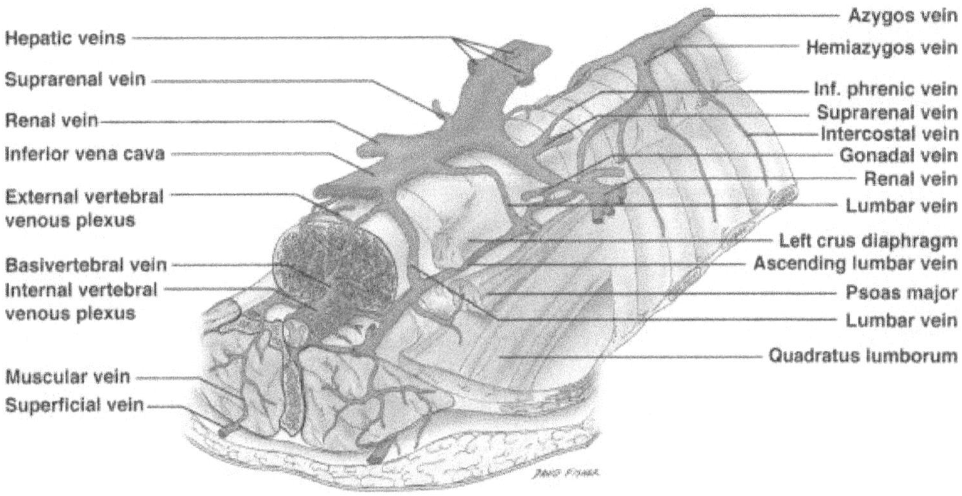

FIGURE 6: Venous drainage

BIOMECHANICS OF THORACOLUMBAR SPINE:

The rigid thoracic spine and the flexible lumbar spine meet at the thoracolumbar junction. The thoracic spine's stiffness, which only permits rotation, is caused by the thoracic muscles, the rib cage, and the facet joints.[21] A sagittal alignment shift occurs between kyphotic thoracic section and the lordotic lumbar region. Each intervertebral segment has roughly 4 degrees of flexion-extension up to T1 and T6. The lateral flexion allowed in the thoracic section is around 8 degrees, but it is approximately 2 degrees[24] in the lumbar spine. This is related to the sagittal orientation of the lumbar spine's facet joints.[22]

PATHOMECHANICS OF INJURY:

Intervertebral mobility in thoracic spine increases significantly when the costovertebral joints are disrupted. Flexion, axial loading, extension, axial rotation and shear are some of the factors at work in spinal trauma.[23] The damage is caused by a mixture of these factors.

Pure axial loads or compressive stresses result in end plate fractures, anterior wedge compression fractures, and burst fractures. The front vertebral body experiences a compressive load from flexion forces with the centre of rotation close to the posterior longitudinal ligament, and the posterior components experience a comparable distraction force. When a sagittal rotation centres at a place in front of the spine, primary distraction forces are applied to both the front and posterior portions of the spine.

Tensile forces are generated in the anterior spine, with the compressive or tensile stresses delivered to the posterior parts in extension-type injuries. The fundamental mechanism causing burst fractures is pure axial force with little extension, resulting in pedicle expansion and fragment retropulsion.[24]

Denis Three column Theory:

To explain the thoracolumbar injury, Denis developed the three-column theory.[25] He cut the spine into three sections. The anterior column is made up of the anterior longitudinal ligament, the front half of the vertebral body, and the annulus with its nucleus pulposus. The middle column is made up of the back half of the vertebral body, the annulus with its pulposis nucleus, and the posterior longitudinal ligament. The neural arch, ligamentum flavum, facet joints, interspinous, and supraspinous ligaments make up the posterior column. Rotation of the dorsolumbar spine and flexion, extension, and lateral bending of the lumbar regions are involved in the movements. The movements become less obvious as people age. The local spinal alignment at the site of damage and the severity both affect the pattern of injury.[26, 27]

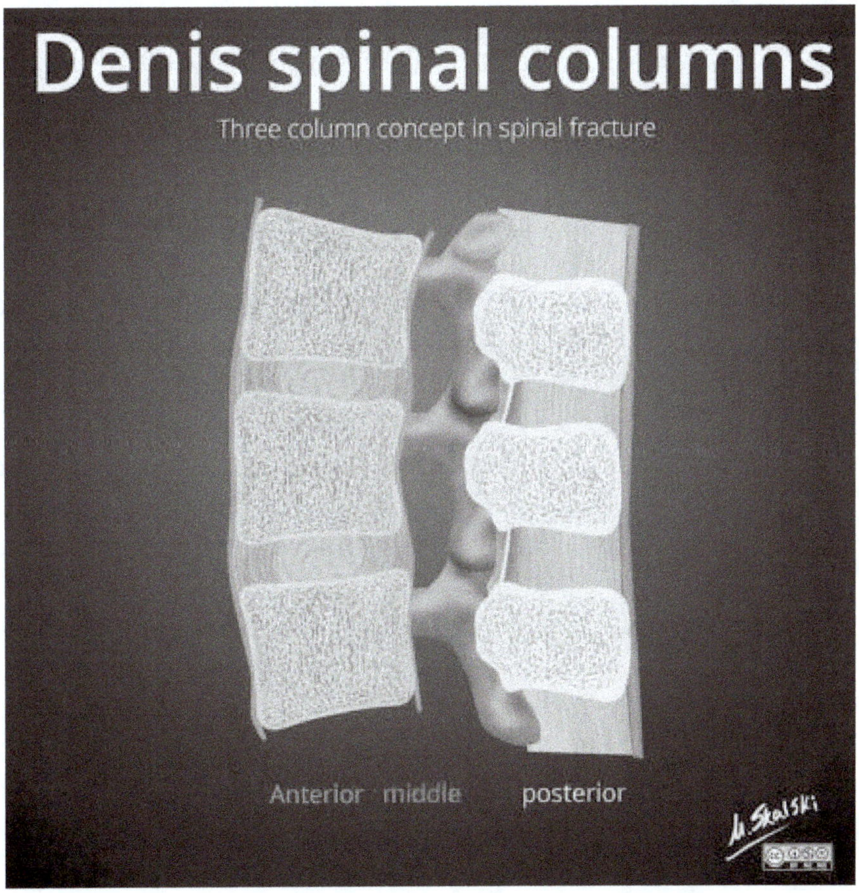

FIGURE 7: Denis 3 columns of spine

CLASSIFICATION OF THORACOLUMBAR FRACTURES

In antiquity, Bohler classified the spinal fractures in 1930.[28] After a series of classification by Watson and Jones' idea of stability & ligamentous integrity in 1949, Holdsworth proposal on the two-column idea in 1963, Louis design on three-column design in 1977, Denis presented the three column theory in 1983, based on CT. [26,29,30,31]

1. Anterior column - anterior longitudinal ligament & anterior half of annulus fibrosus and vertebral body in the Denis anatomical three-column model.
2. Middle column - posterior longitudinal ligament, annulus fibrosus, and vertebral body
3. The posterior column is made up of ligamentous and bony components that are located posterior to the posterior longitudinal ligament.

McCORMACK CLASSIFICATION:

McCormack et al coined the Load sharing classification (LSC) in 1994.[32]

It considers:

1. The amount of comminution in the fractured vertebral body
2. The amount of kyphosis correction needed to restore the sagittal alignment
3. The spread of comminuted fragments in the fractured vertebral body (apposition)

Magrel et al presented Comprehensive classification which consists of:[33]

1. Type A fracture –compression type of injury without posterior lesion
2. Type B fracture - distraction type of injury
3. Type C fracture- rotational type of injury

CLINICAL EVALUATION

The ASIA has suggested that all patients with spinal injuries should have crucial neurological assessments. This includes testing pin prick sensation at 28 particular sensory dermatomes on each side of the body, as well as the motor power of 10 muscles on each side of the body that are innervated by the C5 to T1 and L2 to S1 muscles. The motor and sensory scores are added up, and the results are compared to the norm. The bulbocavernosus response should be studied both at the time of damage and after spinal shock has subsided because it evaluates the majority of the spinal cord.[34, 35, 36]

Following that, a rectal examination should be performed to assess the resting tone, contraction and perianal sensation.

The ASIA suggests the 10 essential muscle groups and nerve roots which are listed below.[35]

Table 1: 10 essential muscle groups and their nerve roots

LEVEL	MUSCLE GROUP
C5	Elbow flexors – Brachialis and Biceps
C6	Wrist extensors – Extensor carpi radialis longus and brevis
C7	Elbow extensors – Triceps
C8	Finger flexors – Flexor digitorum profundus to middle finger
T1	Small finger abductors – Abductor digit minimi
L2	Hip flexors – Illiopsoas
L3	Knee extensors – Quadriceps
L4	Ankle dorsiflexors – Tibialis anterior
L5	Long toe extensors – Extensor hallucis longus
S1	Ankle plantar flexors – Gastrocnemius and Solleus

The functional effects of spinal cord damage are classified as complete or incomplete injuries based on the severity of neurological impairment.

ASIA scores and Frankel's classification classify spinal cord injury patients into five categories.[36]

AMERICAN SPINAL INJURY ASSOCIATION SCORE:[37]

Table 2: ASIA scoring system

A	Complete	No motor or sensory function in the lowest sacral segment
B	Incomplete	Sensory function below neurological level & in S4 S5, no motor function below neurological level
C	Incomplete	Motor function is preserved below the neurological level, key muscle groups below neurological level have a grade < 3
D	Incomplete	Motor function is preserved below the neurological level, key muscle groups below neurological level have a grade >3
E	Normal	Normal motor and sensory function

FRANKEL CLASSIFICATION:[38]

- Grade A: Absent motor and sensory function
- Grade B: Absent motor function, sensory present
- Grade C : Motor function present, but not useful (2 or 3/5), sensory present
- Grade D : Motor function present and useful (4/5), sensory present
- Grade E : Normal motor (4/5) and sensory function

Complete spinal cord injury:

There is no sensation or voluntary movement caudal to the level of damage when the bulbocavernosus reflex is intact below the level of the cord injury, reflex returns.[39]

Incomplete spinal cord injury:

After the bulbocavernosus reflex returns, some neuronal function remains caudal to the level of damage. Presence of perianal sensation, great toe flexor activity all shows sparing of sacral region. It implies partial continuity of white mater tracts with incomplete cord injury, with a higher possibility of cord function recovery following spinal shock resolution. Better the prognosis, greater the function distal to injury and the faster the recovery.[40]
The most common type of cord injury is a spinal contusion. Because of inflammation and bleeding surrounding the injury, the cord is just contused but not cut. It results in a temporary shutdown of cord functions. It usually resolves itself within 24-48 hours.

Anterior cord syndrome:

Damage to the anterior portions of the spinal cord causes complete motor loss, loss of warmth, and discomfort. It is particularly common in hyperflexion injuries where bone or disc fragments compress the anterior spinal artery. Deep touch, location sense, and vibratory feeling are all maintained. The outlook for recovery is bleak.[41]

Posterior cord syndrome:

It frequently happens in injuries of the extension form. The cord's posterior column is affected. Following is the loss of location sense, vibration sense, and deep touch. This condition is hardly frequent.[42]

Brown-Sequard syndrome:

Movement & sensation are lost on the same side, while pain, body temperature are lost on the opposite side below the degree of injury. It happens when the right/left side of spinal cord is injured.[43]

Conus medullaris syndrome:

Since the cord ends at L1, this syndrome develops when a damage occurs between T11 and L2. As a result, damage to the lumbar nerve roots or conus occurs. It causes motor loss in the lower extremities as well as areflexia of the bladder and colon. Perianal wink and the bulbocavernous reflex won't be present.[44]

Cauda equina syndrome:

It is a less severe and partial motor neuron injury. It occurs when there is damage between the conus and the lumbosacral nerve roots and beyond the L1 vertebra. It results in the absence of the bulbocavernosus reflex, the anal wink, and all other forms of reflex activity in the lower limbs, as well as areflexic bladder, bowel, and asymmetrical motor and sensory loss.[44]

Mixed syndrome:

Typically, it combines many disorders. It refers to small proportion of incomplete spinal cord injuries that do not meet one of the symptoms that have been previously mentioned.

Grading of spinal cord injury:

For grading spinal cord damage, the Frankel method and the American Spinal Injury Association (ASIA) are often employed. It is crucial in assessing therapy and long-term prognosis for many patients who have suffered thoracolumbar spinal damage.[45]

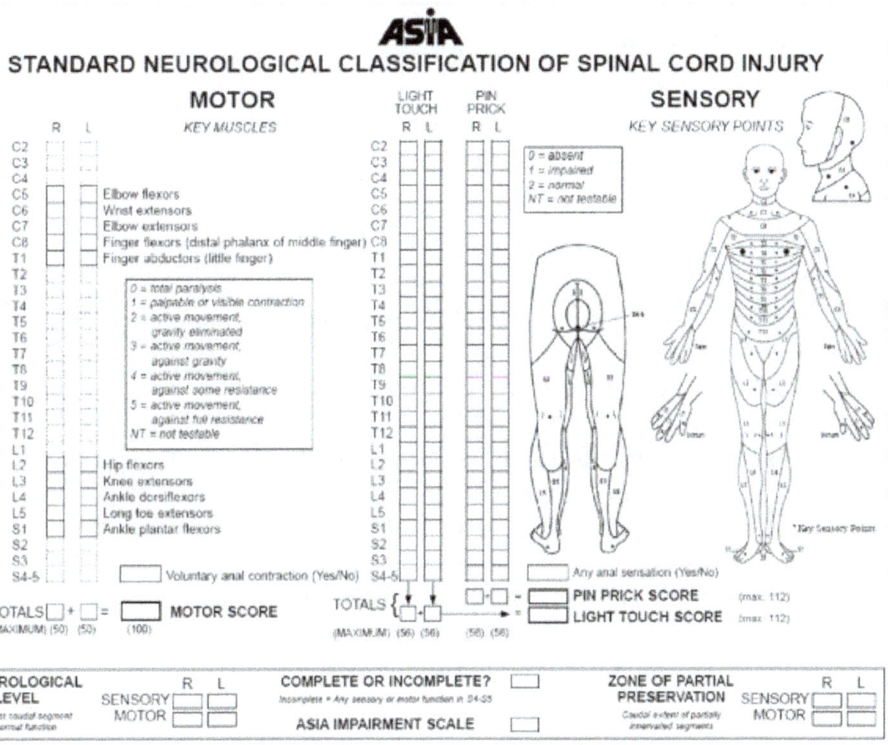

FIGURE 8 : ASIA charting

RADIOGRAPHIC-EVALUATION

X-RAYS:

Anteroposterior & lateral radiographs of cervical, thoracic, and lumbar spine radiographs should be obtained. Because of high occurrence of contiguous and noncontiguous related spinal fractures, a thorough examination is indicated for every patient who has received a high-energy injury and is suspected of having a spinal injury. In AP view, we must consider interpedicular and interspinous distances. Coronal translational deformation must be investigated. The posterior vertebral line, sagittal malalignment, vertebral body height, and Cobbs angle should be seen in lateral view.[46]

COMPUTED TOMOGRAPHY:

Bony injuries will be precisely described by computed tomography. CT scanning comes next after the examination using plain radiography. Sagittal sections and axial fine slices are used to understand the type of fracture and degree of spinal canal injury. It is important to measure the canal diameter of fractured vertebra. It will give details on vertebral body fragmentation.[47]

MAGNETIC RESONANCE IMAGING:

MRI is critical for determining cord & ligamentous complex involvement. It provides detail on canal compromise, cord alterations & epidural hematomas. It also will provide information on the integrity of the PLC. It is critical in determining treatment. A gas shadow implies flexion distraction injury. However, when a patient has a neurological disability, MRI is advised to detect any persistent spinal cord compression.[48]

TREATMENT:

The goal of treatment in spinal fractures is to mobilise the patient as soon as possible and to achieve a stable vertebral column with maximum spine motion & good neurological status. General condition, neurological health, spinal stability, degree of deformity, and accompanying injuries are the most significant things to consider when treating patients. Recently, the Vaccaro score has proven to be extremely beneficial in determining treatment.[49]

THORACOLUMBAR INJURY CLASSIFICATION AND SEVERITY SCORE (TLICS SCORE, VACCARO SCORE):[50, 51]

FRACTURE MECHANISM	
Compression fracture	1
Burst	2
Translation3	3
Distraction	4
NEUROLOGICAL INVOLEMENT	
Intact	0
Nerve root	2
Cord,conus medullaris,incomplete	3
Cord,conus medllaris,complete	2
Cauda equine	3
POSTERIOR LIGAMENTOUS COMPLEX INTEGRITY	
Intact	0
Injury suspected/indeterminate	2
Injured	3

Table 3: TLICS scoring system

1. Score less than 3 –it is a stable fracture so it can be treated conservatively
2. Score 5 or more –it is a unstable one. So it should be treated surgically.
3. Score 4- decision by the surgeon

MANAGEMENT

Conservative management:

Non-operative treatment was being utilised for spine fractures until 1970. Distraction, reduction, and rest on scamnum were used to treat the patients (low bench). Bed rest, postural reduction, direct mobilisation, ambulatory bracing, and a combination of these were used.

The majority of compression fractures are stable wounds that are treated non-operatively for twelve weeks while wearing a thoracolumbar orthosis. When the canal is damaged, burst fractures become more unstable. The degree of vertebral body collapse, the kyphotic angle, and the integrity of the posterior column are crucial factors in deciding whether surgery or non-surgical treatment is appropriate. Surgical decompression and stabilisation are frequently required in cases of neurological degeneration with substantial canal damage. When injuries are stabilised and the neurology is undamaged, conservative measures can be performed.[52]

Surgical management:

Several elements such as age, health status, occupation, lifestyle, and body habitus are all important factors in achieving a positive outcome. Due to high intensity trauma, the majority of the young patients require extensive surgical therapy.[53]

Dorsolumbar spine fractures can be treated surgically in three ways. Anterior, posterior and combined surgical approaches are available. Most cases of compression fractures & burst fractures involving the anterior & middle columns can be treated with anterior instrumentation without involving posterior column. However, surgical complications are common with this approach. Morbidity and mortality are also high.

A posterior technique using instrumentation from two or three levels cephalad to the damage to two levels caudad (long segment fixation) is a suitable alternative treatment, however it requires fusing a greater section of the spine.[54]

A combined anteroposterior fusion frequently offers the highest level of iatrogenic spinal stability when three columns are involved. Fractures, dislocations & flexion-distraction types are best treated from the back initially, with posterior stabilisation.[55]

If there is full neurological impairment, surgery is advised to achieve immediate spinal stabilisation and to allow for faster recuperation. In this scenario, the posterior approach is suitable for restoring stability.

Neural Decompression:[56]

There are several types of decompression.

1. Direct anterior decompression
2. Posterior decompression
    a) Indirect posterior decompression by ligamentotaxis
    b) Postero lateral decompression (transpedicular, costo transversectomy)
    c) Laminectomy

When there is neural compression and neurological loss due to spine fractures, spinal canal decompression is required. Although the extent of canal compromise at the time of the initial evaluation has been linked to the severity of the neurological deficit in certain studies, the timing of surgical decompression has not been linked to the degree of neurological recovery.

However, most researchers agree that severe neurological deterioration warrants surgical intervention. When there is evidence of persistent compression of the neural components, for a patient with a stable incomplete neurological impairment, many surgeons consider surgical decompression. Because the source of the neural compression is usually found anteriorly in most patients, an anterior approach is preferable.[57]

By using distraction instruments and ligamentotaxis, a posterior surgical method may occasionally be used to indirectly decompress the spinal canal. The effectiveness of a posterior indirect decompression's ligamentotaxis depends on the sharpey fibre or annular ligament attachments to the dislocated fracture fragments. Utilizing posterior instrumentation, the distraction force will reduce the fragment from the canal into the vertebral body.[58]

Sometimes posterior distraction exacerbates any inherent kyphosis. When surgery is performed within 72 hours of the traumatic incident, the efficacy of indirect decompression is increased. This procedure may be less effective in cases when the canal compromise is greater than 70%, as this is associated with higher prevalence of annular ligament disruption.

Stabilization :

The majority of unstable thoracolumbar fractures require surgery to stabilise them. A stable spine is desired in order to facilitate early mobilisation and early recovery.

1. Long segment spanning fixation
2. Short segment with index vertebra fixation

Early surgical stabilisation of unstable fractures has the benefits of improved fracture reduction, preservation of neurological function, early mobilisation, and a reduced risk of sequelae from prolonged bed rest. In order to limit the amount of motion segments that needed to be fused, long with short fusion techniques were devised. It entails employing pedicle screws to cross the cephalad and caudad levels while fusing the two levels that are close to the damaged vertebra. However, due to arthritic changes in segments which are unfused and growing kyphosis after the removal of the rod and the pain returning, the results were poor. Due to this, fixation without fusion is advised as a routine in place of this procedure, which has lost favour. More dependable stiff fixation is made possible by contemporary pedicle screw-based systems. A method that offers a large amount of stability is long segment spanning fixation, which entails fixing two levels above and below the fracture location. A more modern technique to keep the motion in place is short-segment fixation with pedicle screw instrumentation (from a level cephalad to the fracture site to a level caudad to it). Enhancing intraoperative deformity reduction, promoting fracture healing, maintaining the reduction, and being associated with low failure rates are all benefits of incorporating a brief segment with index vertebra fixation and the fractured vertebra.[59, 60, 61]

Review:

In a research by Jun Li (2016) that covered 455 cases, there was no tangible difference in the clinical result between short and long segment fixation (239 short segment and 216 long segment).

Long segment fixation required more time during surgery, but the results were good radiologically, and there was less implant failure.[62]

Four fixation models were employed in a study by Yue zhou and Changing li (2014), and it was determined that adding pedicle screws closer to the fracture may make the construct sturdier and cause less stress on the pedicle screws and rods.[63]

18 consecutive patients were involved in a study by Tezeren G. Kuru (2005), with 9 of them receiving short segment therapy and 9 receiving long segment therapy. Local kyphosis, sagittal index & vertebral body height compression measurements revealed that long segment fixation had better results at the end of the treatment. Additionally, there was a 55% failure rate in the short segment group, but there was a longer operating time and more blood loss in the long segment group. According to the low back outcome score, there was no significant difference between the two groups.[64]

A safe and effective treatment option for thoracolumbar fractures is posterior short segment pedicle screw fixation with pedicle screw at the fracture level, according to many studies. This may help to maintain and/or lessen kyphosis.[65]

In a research on the indirect reduction and posterior stabilisation of unstable thoracolumbar burst fractures by George et al stated that both long-segment and short-segment stabilisation can reduce segmental kyphosis and vertebral body deformation after surgery. However, over time, long-segment stability has been linked to improved outcomes in terms of the radiological characteristics, indices, and patient satisfaction.[66]

Long-segment stabilisation was found to be advantageous in terms of minimising overall stress on the spine in Recep et al finite element analysis of short- versus long-segment posterior fixation for thoracolumbar burst fracture. However, by including the spinal fracture in the stabilising system, the strain on the system could be lessened. So as long as nearby undamaged vertebrae are maintained, short-segment stabilisation is adequate to establish a secure and reliable stability system.[67]

Ambrose et al assessed the radiological outcome of short segment posterior stabilisation and fusion in thoracolumbar spine fracture. Between the patients' preoperative, postoperative, and follow-up sagittal index, anterior body compression, and regional kyphosis measurements, a statistically significant difference was observed.[68]

MATERIALS AND METHODS

This study was conducted between October 2020 and 2022. The patients presented to the emergency room and were assessed and treated according to the ATLS protocol. All the patients received a thorough clinical and neurological examination. The level of the injury sustained by the patient was diagnosed with clinical and radiological correlation according to AO spoine trauma classification system. The injuries sustained by the patients were classified according to ASIA scoring system and TLICS scores were calculated for all the patients in this study. Cobb's angle and beck's index were also calculated pre-operatively for all patients. A total of 30 patients were a part of this study with 15 patients getting a short posterior stabilization and the other 15 getting long posterior stabilization.

INCLUSION CRITERIA

- Patients willing to enroll in this study for surgical procedure and follow-up
- Age 20 - 80
- TLICS more than or equal to 4
- Single level thorcolumbar vetrebral burst fracture

EXCLUSION CRITERIA

- Pathological fracture
- Metabolic bone disease
- TLICS score less than 4
- Pregnant patients
- Patients who were not willing to get enrolled.

The neurological deficit was quantified according to ASIA grading. Radiographs were taken followed by CT and MRI scans to assess the retropulsion of the fractured vertebra, canal compromise, status of the cord and integrity of the posterior longitudinal ligament. An E-FAST was also done and general surgery clearance was obtained. These patients underwent either short or long segment posterior instrumentation. The decision to proceed with either short or long segment posterior stabilisation was based on convenience sampling.

OPERATIVE TECHNIQUES

General anethesia was administered by the anesthesisa team following which the patient was positioned prone on the surgical table. The patient was put in prone position with bolsters at appropriate sites to keep the abdomen free from tension/pressure. The level of injury was identified and confirmed under c-arm guidance. The incision was made through standard posterior approach. The spinous processes of the vertebrae were indentified and the surgical plane between the spinous process and the paraspinal muscles was made. The paraspinal muscles were erased by using a cobb's spinal elevator, and subsequently were reflected laterally with self retaining retractor. Pedicles were identified by the intersecting point between middle of the transverse process and the longitudinal axis of the superior facet.

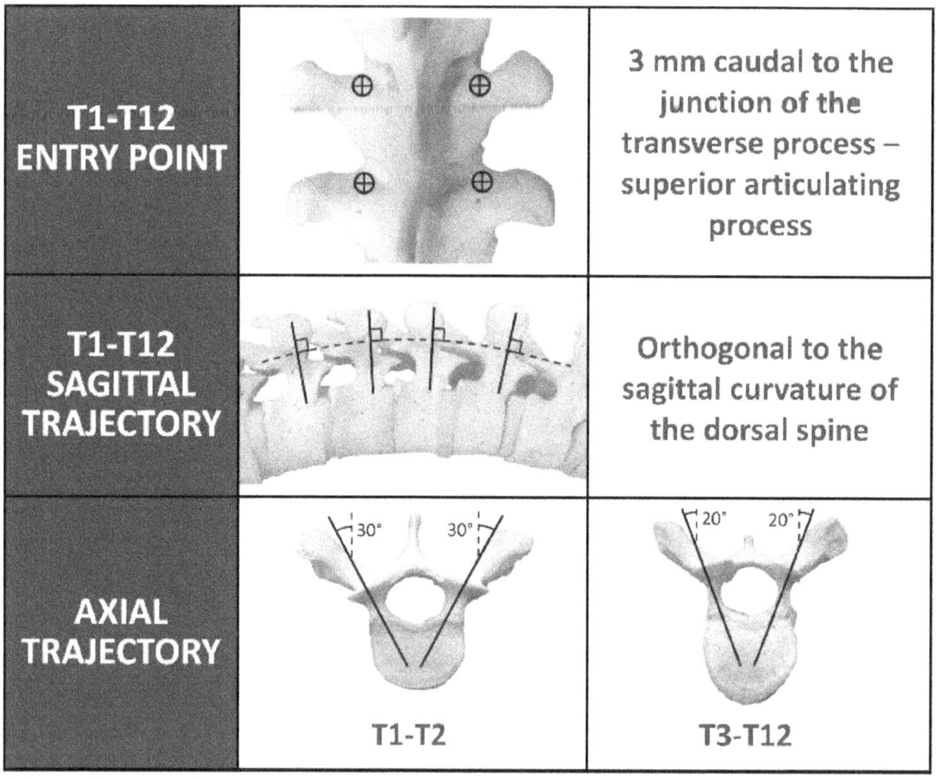

FIGURE 9: Dorsal spine entry point for pedicle screws

Lumbar pedicle screw entry points

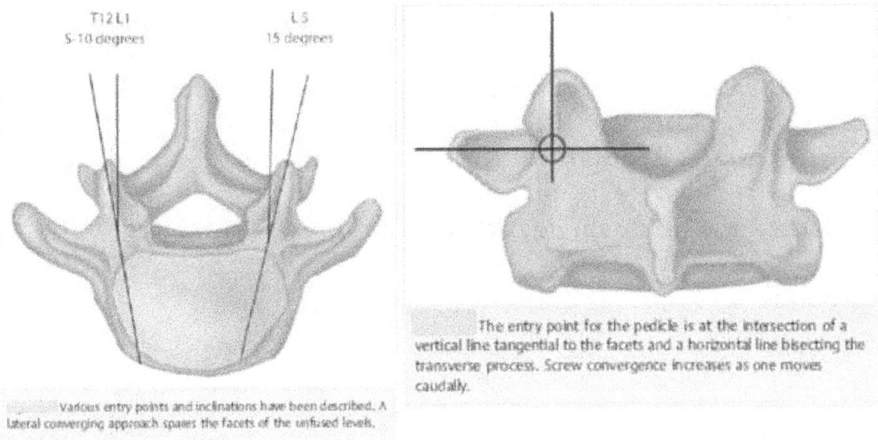

FIGURE 10: Entry point at DL junction and lumbar spine

LUMBAR SPINE: ENTRY POINTS

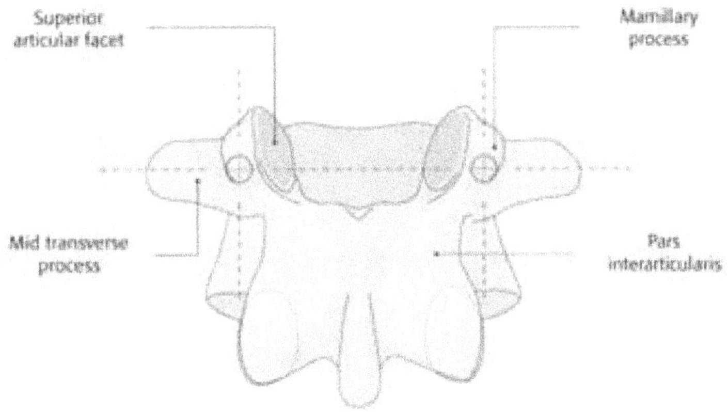

Figure 11: Entry point for lumbar spine pedicle screws

The pedicle screws were inserted under image intensifier guidance after probing and measuring the depth of the pedicles. Appropriate sized screws were used in the dorsal and lumbar pedicles. Implants used were from Jayon medical devices pvt ltd.

Thorough decompression was done by doing a laminectomy at the fracture site,both cords and roots were confirmed to be free. Connecting rods were placed and fixed with screws. Horizontal cross links were added if needed. Intertransverse fusion was done with bone graft obtained from the remnants of the decompression. Closure was done in layers and drain was inserted with negative suction.

Instruments and implants images

FIGURE 12: Pedicle screws and connecting rods

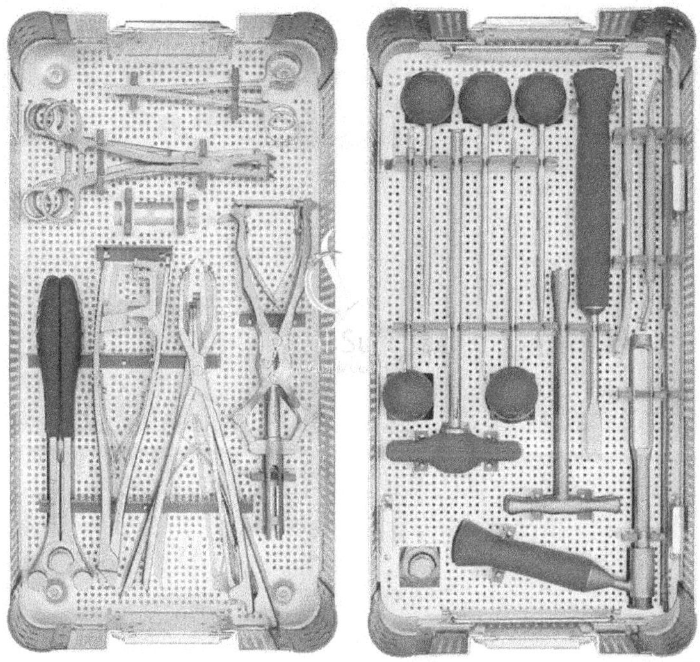

FIGURE 13: Surgical spine set

Intra-op images

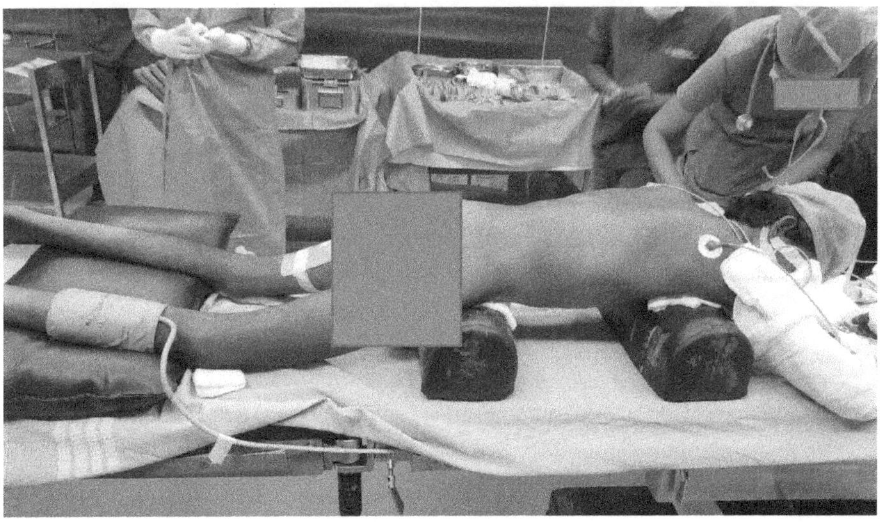

FIGURE 14. 1: Positioning

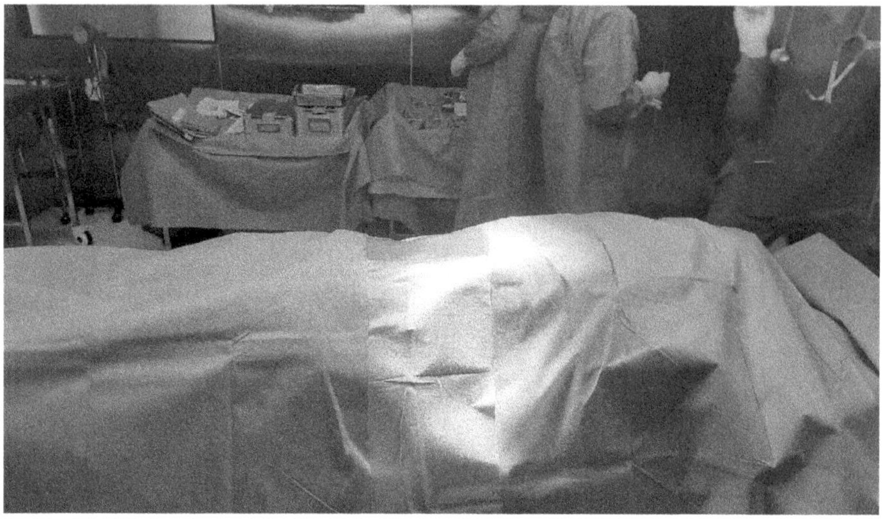

FIGURE 14. 2 After draping

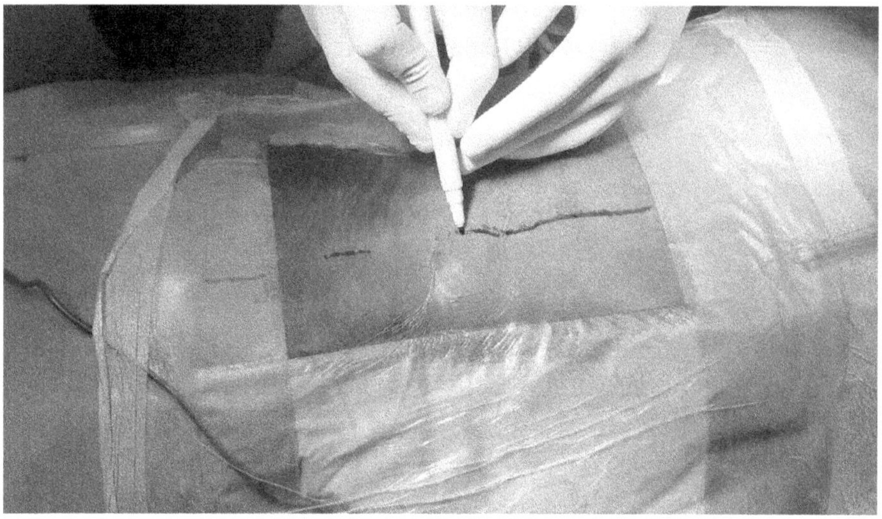

FIGURE 14. 3 Marking of incision site

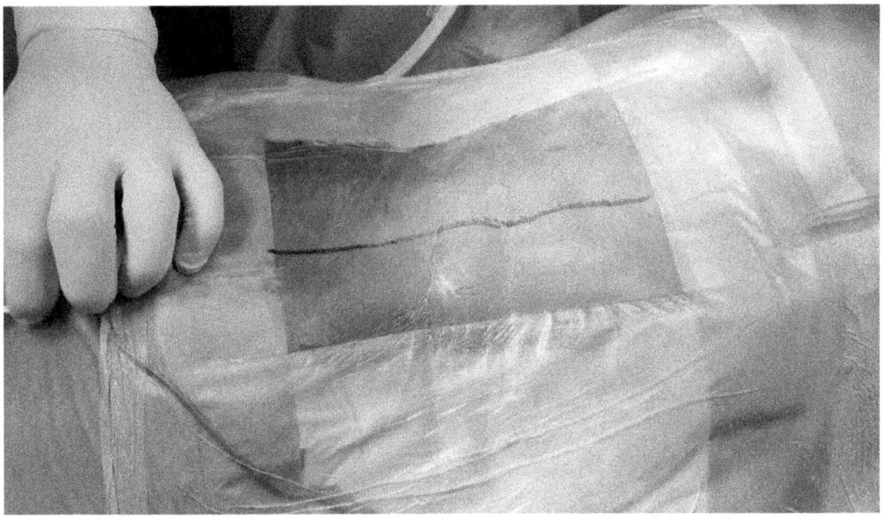

FIGURE 14. 4 Incision site

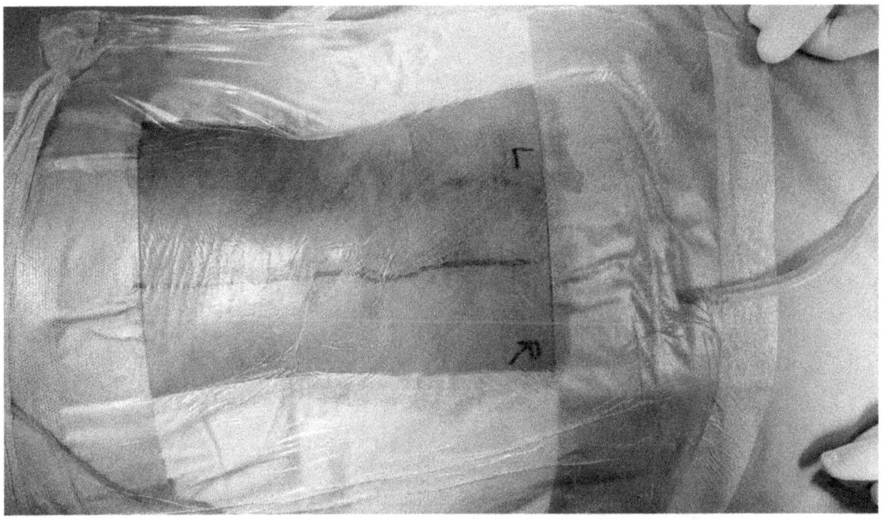

Figure 14. 5 Incision site

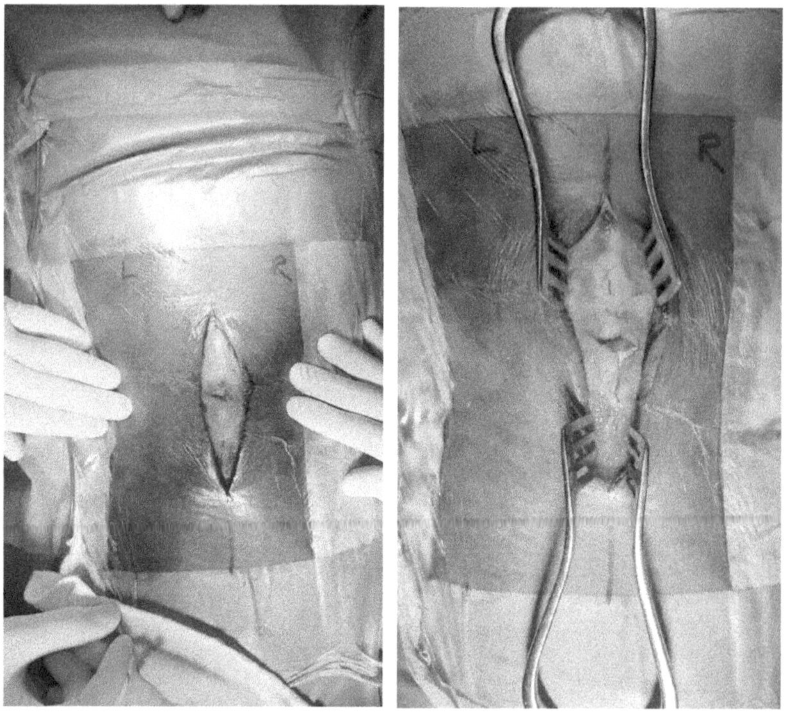

FIGURE 14. 6

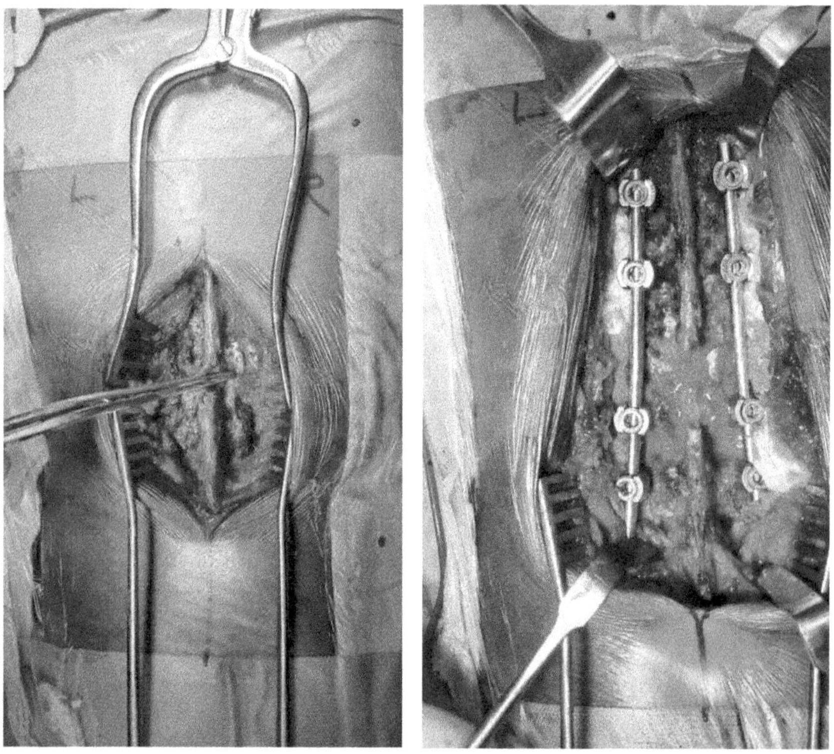

FIGURE 14. 7

Postoperative protocol

Patients were provided with in-bed mobilization in the postop period. On postoperative day 2 drain was removed. Dressings were done on postoperative day 2 and 5. Suture removal was done on postoperative day 12. Mobilization was started as per patient's tolerance. Stage appropriate neurorehabilitation was provided by the physiotherapists. Post-op x-rays were taken on postoperative day 1-2.

Follow-up

The patients were followed up every month till 6 months and every 2 months till 1 year. During the follow-up period, patients were assessed using Denis pain and work assessment scale.

DENIS PAIN AND WORK SCALE:

Pain Score	
P1	No pain
P2	Occasional minimal pain, no need for medication.
P3	Moderate pain, occasional medications and no interruption of work or activities of daily living.
P4	Moderate to severe pain, occasionally absent from work, significant changes of activities of daily living.
P5	Constant severe pain, chronic pain medications.
Work score	
W1	Return to previous employment (heavy labor) or physically demanding activities.
W2	Able to return to previous employment (sedentary) or return to heavy labor with restrictions.
W3	Unable to return to previous employment but works full time at new job.
W4	Unable to return to full time work.
W5	No work completely disabled.

Table 4: Denis

Pain Score	
P1	No pain
P2	Occasional minimal pain, no need for medication.
P3	Moderate pain, occasional medications and no interruption of work or activities of daily living.
P4	Moderate to severe pain, occasionally absent from work, significant changes of activities of daily living.
P5	Constant severe pain, chronic pain medications.
Work score	
W1	Return to previous employment (heavy labor) or physically demanding activities.
W2	Able to return to previous employment (sedentary) or return to heavy labor with restrictions.
W3	Unable to return to previous employment but works full time at new job.
W4	Unable to return to full time work.
W5	No work completely disabled.

STATISTICAL ANALYSIS

The data obtained was charted using MS-excel [microsoft USA] and analysed using SPSS version 16[IBM]. The numerical variables such as age, beck's index and cobb's angle were summarised as mean and standard deviation, while the categorical variables such as gender, mode of injury, diagnosis and ASIA impairment scale, were categorised as frequency and percentage. The comparison between the independent and dependent variables were done using independent t test for numerical variables and chi-square test for categorical variables. The timelines within each group were compared using repeated measures of analysis of variance (RMNOVA). The graphical distribution of the variables was done using pie charts, bar charts and box and whisker plots. All p-values less than 0. 05 were considered significant.

RESULTS

Table 5: Intervention distribution among the patients (n=30)

		Number	Percentage
Group	Long segment stabilization	15	50%
	Short segment stabilization	15	50%

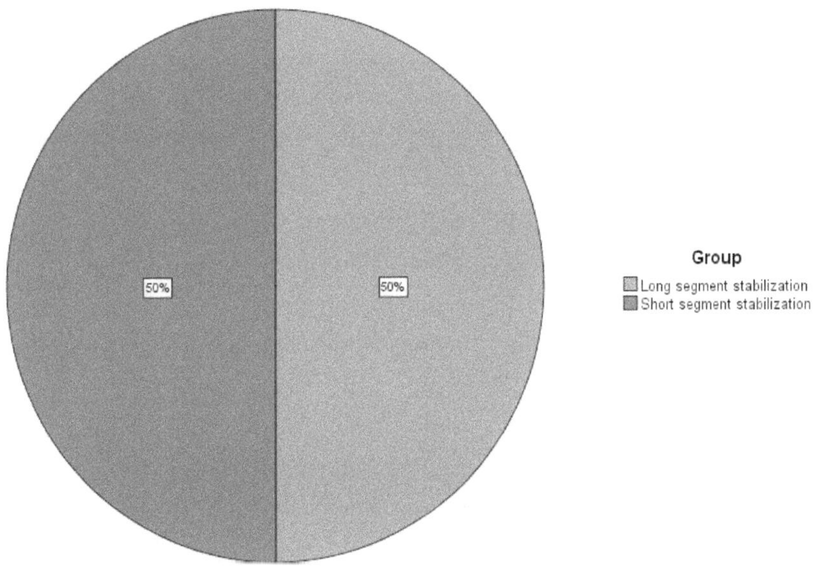

Figure 15: Graphical representation of the distribution of intervention among the patients (n=30)

Among the 30-paticipants of the study, exactly half of the patients underwent long segment and the remaining underwent short segment stabilization.

Table 6: Age distribution among the intervention groups (n=30)

		Group		p value*
		Long segment stabilization (n=15)	Short segment stabilization (n=15)	
Age	Mean	38	35	0.517
	Standard Deviation	17.0	12.7	
	Median	38	35	
	Percentile 25	27	22	
	Percentile 75	46	45	

**p value by independent t test*

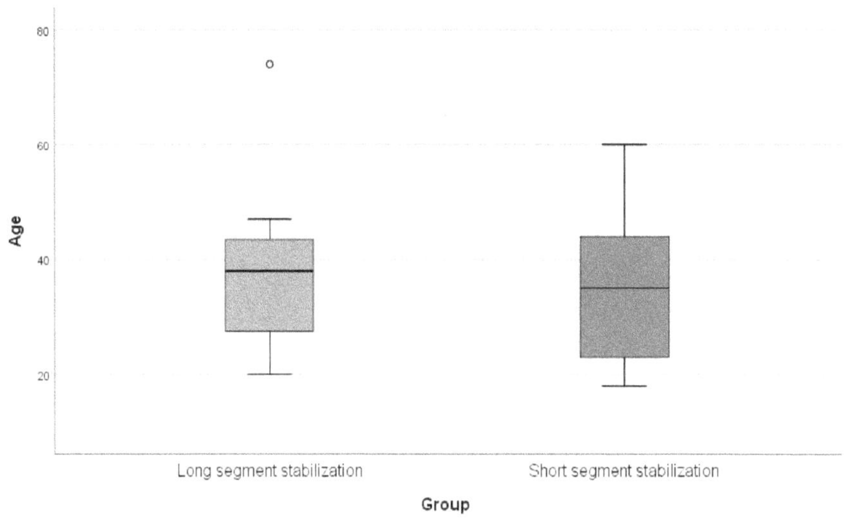

Figure 16: Graphical distribution of the Age distribution among the intervention groups (n=30)

The mean age of the participants was 38 (±17) years in the long segment group and 35 (±12.7) years in the short segment group. The comparison of the age was statistically insignificant inferring equal distribution in both groups.

Table 7: Age category distribution among the intervention groups (n=30)

		Group, n (%)		p value*
		Long segment stabilization (n=15)	Short segment stabilization (n=15)	
Age category	≤30 years	7 (46.7%)	6 (40.0%)	0.734
	31- 45 years	5 (27.8%)	6 (40.0%)	
	> 45 years	4 (26.7%)	3 (20.0%)	

* *p value by chi-square test*

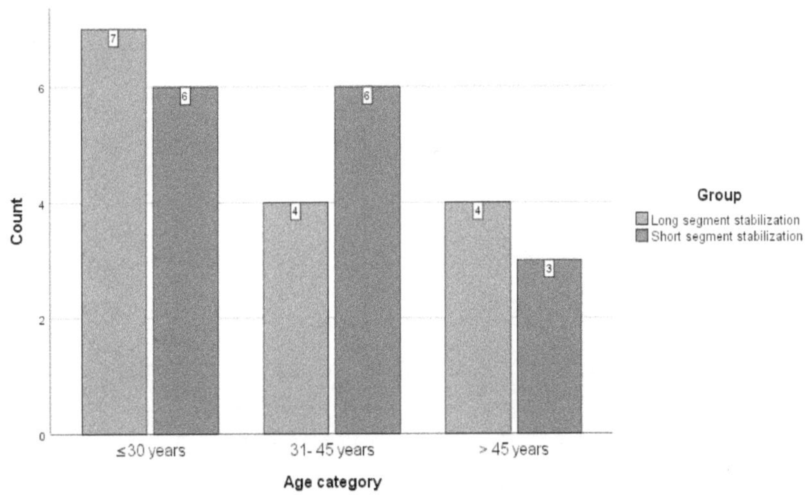

Figure 17: Graphical distribution of the Age category distribution among the intervention groups (n=30)

Among the two interventional group of patients, majority (47%) were less than 30 years old in the long segment group while in the short segment group, there was equal distribution of patients in the less than 30 years and 30-45-year age groups. The comparison of the age category was statistically insignificant inferring equal distribution in both groups.

Table 8: Gender distribution among the intervention groups (n=30)

		Group, n(%)		p value*
		Long segment stabilization (n=15)	Short segment stabilization (n=15)	
Gender	Male	11 (73.3%)	11 (73.3%)	1.000
	Female	4 (26.7%)	4 (26.7%)	

** p value by chi-square test*

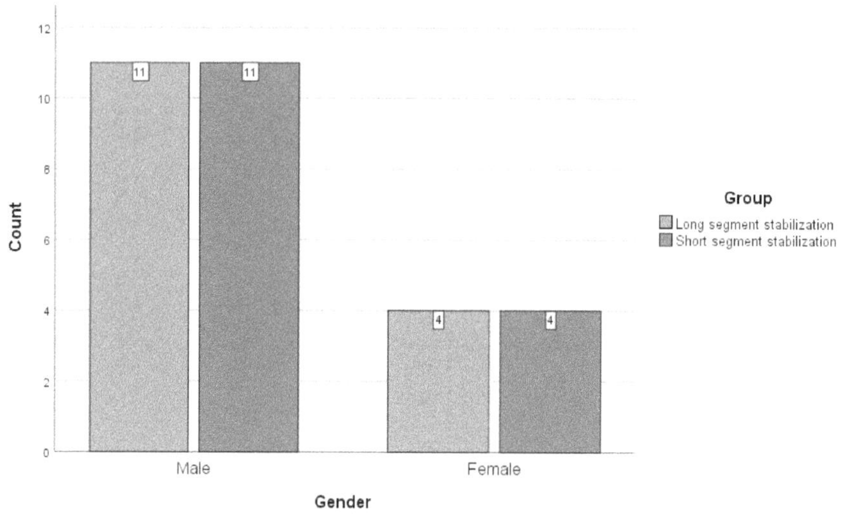

Figure 18: Graphical representation of the gender distribution among the intervention groups (n=30)

Most of the patients (73%) in both the intervention groups were males. The comparison of the gender was statistically insignificant inferring equal distribution in both groups.

Table 9: Mode of injury among the intervention groups (n=30)

		Group, n(%)		p value*
		Long segment stabilization (n=15)	Short segment stabilization (n=15)	
Mode of injury	RTA	9 (60.0%)	7 (46.7%)	0.464
	Fall from height	6 (40.0%)	8 (53.3%)	

** p value by chi-square test*

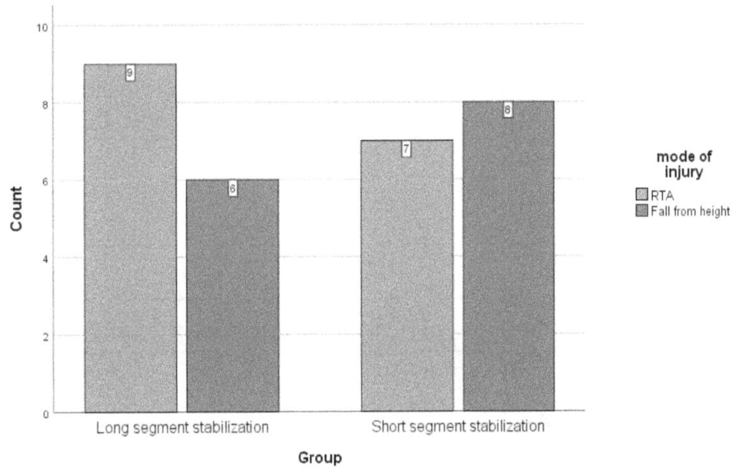

Figure 19: Graphical representation of Mode of injury among the intervention groups (n=30)

For the mode of injury, majority (60%) were due to RTA in the long segment group while more than half (53%) in the short segment group were from fall from height. However, the comparison of the mode of injury was statistically insignificant inferring equal distribution in both groups.

Table 10: Comparison of diagnosis and levels fused among the intervention groups (n=30)

		Group, n(%)	
		Long segment stabilization (n=15)	Short segment stabilization (n=15)
Diagnosis	D10	0	1 (6.7%)
	D12	3 (20.0%)	4 (26.7%)
	L1	8 (53.3%)	7 (46.7%)
	L2	1 (6.7%)	1 (6.7%)
	L3	3 (20%)	2 (13.3%)
Level fused	D9 - D11	0	1 (6.7%)
	D10 - L2	3 (20%)	0
	D10 - L3	3 (20%)	0
	D11 - L1	0	4 (26.7%)
	D11 - L3	5 (43.3%)	0
	D12 - L2	0	7 (46.7%)
	D12 - L4	1 (6.7%)	0
	L1 - L5	3 (20.0%)	0
	L1- L3	0	1 (6.7%)
	L2 - L4	0	2 (13.3%)

Among the patients, majority were having L3 fracture in both long segment (53%) and short segment groups (47%). Similarly, most of the patients (43%) in the long segment had levels fused at D11-L3 while in the short segment group most of them (47%) had fusion from D12-L2.

Table 11: TLICS comparison among the intervention groups (n=30)

		Group, n(%)		p value*
		Long segment stabilization (n=15)	Short segment stabilization (n=15)	
TLICS	4	2 (13.3%)	2 (13.3%)	0.328
	6	9 (60.0%)	12 (80.0%)	
	7	4 (26.7%)	1 (6.7%)	

** p value by chi-square test*

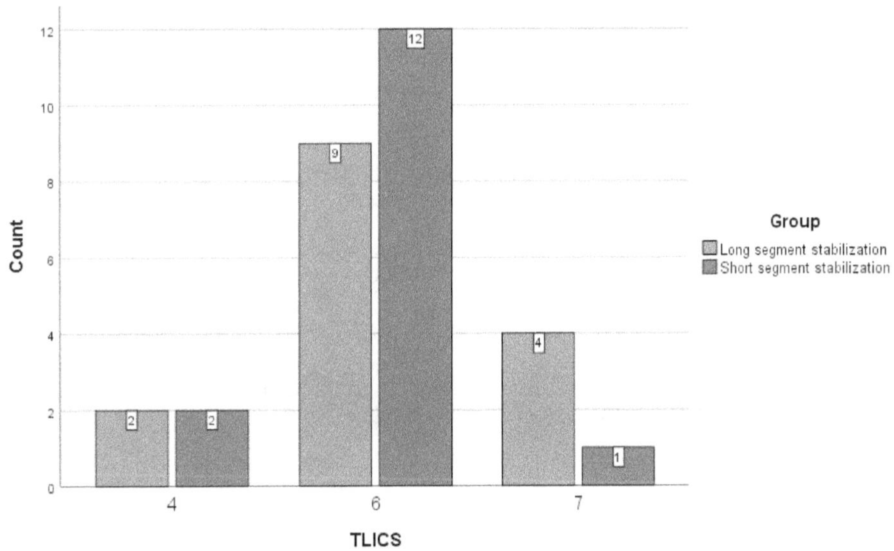

Figure 20: Graphical representation of TLICS comparison among the intervention groups (n=30)

The TLICS comparison between the two groups revealed that majority had a score of 6 in both the long (60%) and short segment (80%) stabilization. The comparison of the TLICS was statistically insignificant inferring equal distribution in both groups.

Table 12: Comparison of operating time among the intervention groups (n=30)

		Group		p value*
		Long segment stabilization (n=15)	Short segment stabilization (n=15)	
Operating time (minutes)	Mean	136.1	79.4	<0.001#
	Standard Deviation	11.31	11.70	
	Median	134	77	
	Percentile 25	128	69	
	Percentile 75	144	86	

**p value by independent t test: #-p<0.05*

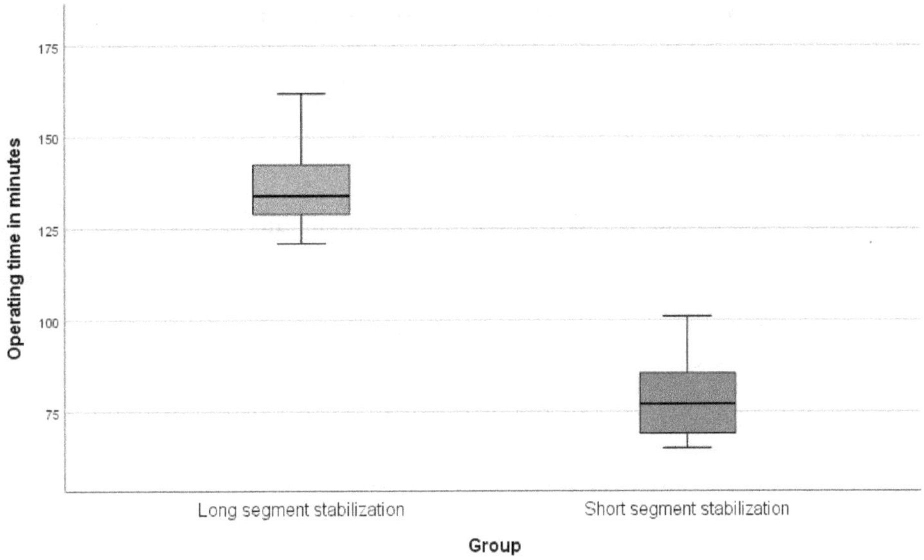

Figure 21: Graphical representation of comparison of operating time among the intervention groups (n=30)

The operating time in the long segment and the short segment groups were 136.1 (±11.31) and 79.4 (±11.7) minutes respectively. The difference between the groups was statistically significant ($p<0.05$) inferring that the time taken to operate was significantly less in the short segment group on comparison with the long segment group.

Table 13: Comparison of blood loss among the intervention groups (n=30)

		Group		p value*
		Long segment stabilization (n=15)	Short segment stabilization (n=15)	
Blood loss (ml)	Mean	1263.3	876.7	<0.001#
	Standard Deviation	151.74	189.80	
	Median	1250	800	
	Percentile 25	1200	700	
	Percentile 75	1400	1100	

**p value by independent t test: #-p<0.05*

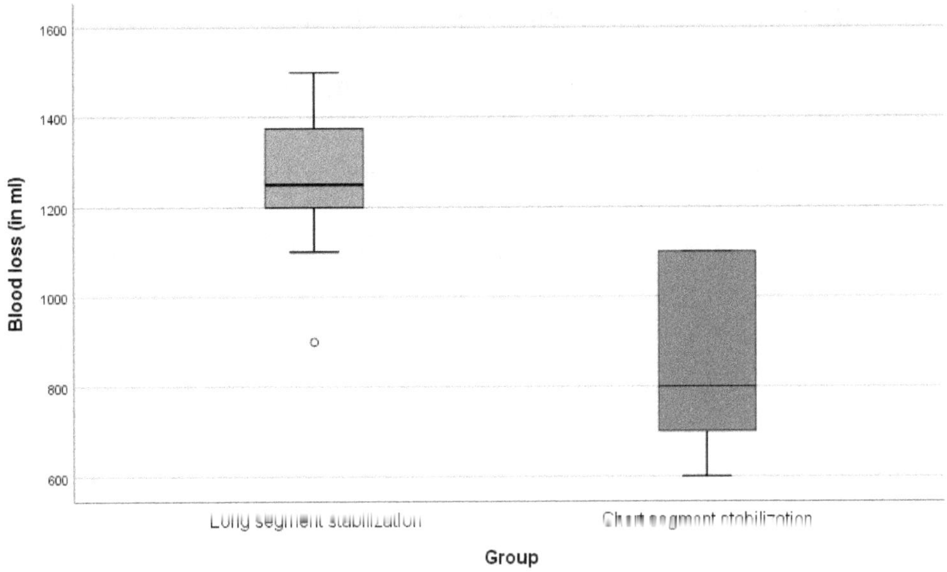

Figure 22: Graphical representation of Comparison of blood loss among the intervention groups (n=30)

The blood loss in the long segment and the short segment groups were 1263.3 (±151.74) and 876.7 (±189.8) minutes respectively. The difference between the groups was statistically significant ($p<0.05$) inferring that the blood loss was significantly less in the short segment group on comparison with the long segment group.

Table 14: ASIA impairment scale at different timelines among the intervention groups (n=30)

		Group, n(%)		p value
		Long segment stabilization (n=15)	Short segment stabilization (n=15)	
ASIA impairment scale- Preoperative	C	8 (53.3%)	6 (40.0%)	0.751
	D	6 (40.0%)	8 (53.3%)	
	E	1 (6.7%)	1 (6.7%)	
ASIA impairment scale at 3 months	C	5 (33.3%)	6 (40.0%)	0.809
	D	8 (53.3%)	8 (53.3%)	
	E	2 (13.3%)	1 (6.7%)	
ASIA impairment scale at 6 months	C	2 (13.3%)	5 (33.3%)	0.297
	D	10 (66.7%)	6 (40.0%)	
	E	3 (20.0%)	4 (26.7%)	
ASIA impairment scale at 1 year	C	1 (6.7%)	5 (33.3%)	0.135
	D	8 (53.3%)	4 (26.7%)	
	E	6 (40.0%)	6 (40.0%)	

** p value by chi-square test*

The above table shows the comparison of the ASIA impairment scale in the preoperative period and at 3, 6 and 12 months. Majority of the patients in the long segment group and short segment group were in class D at all timelines. The difference between the two groups was statistically insignificant at all timelines inferring equal distribution in both groups.

Figure 23: Graphical representation of the ASIA impairment scale at different timelines among the intervention groups (n=30)

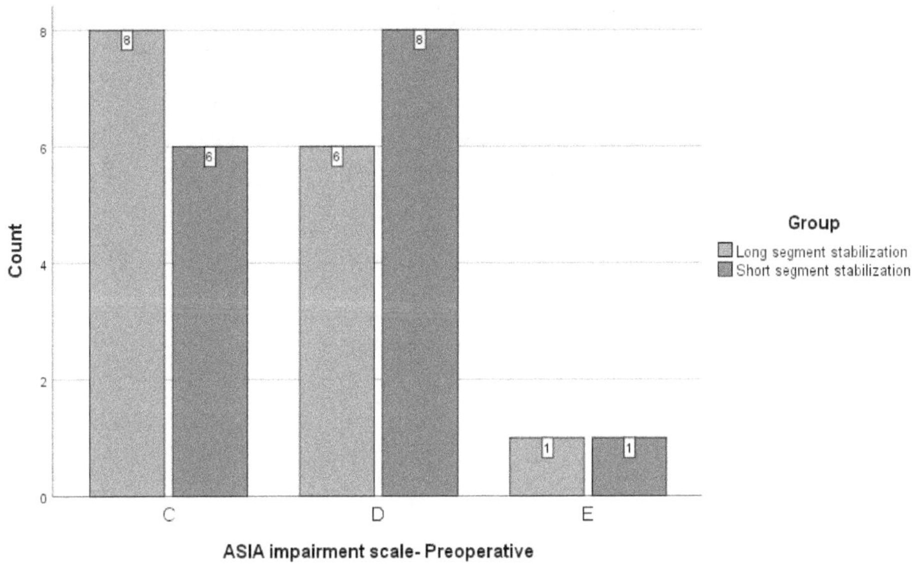

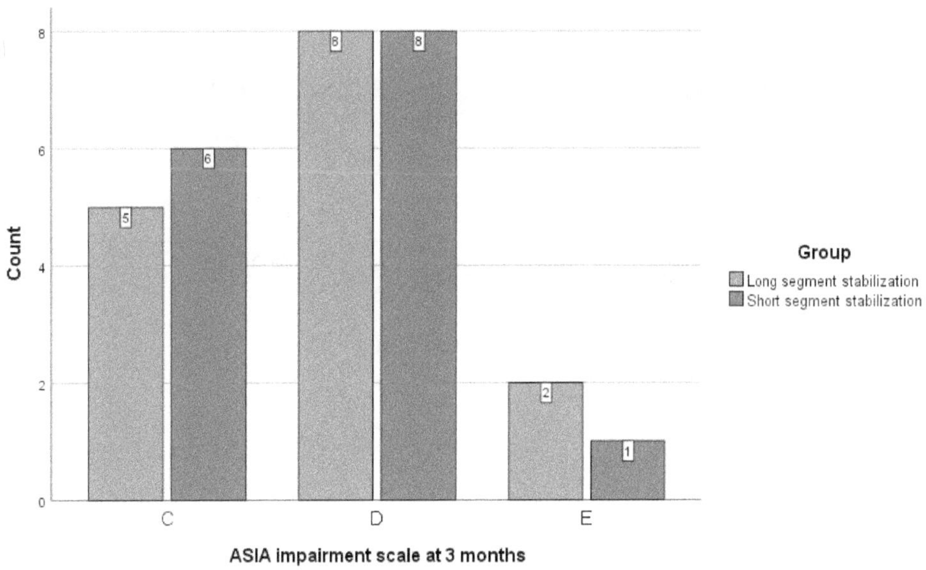

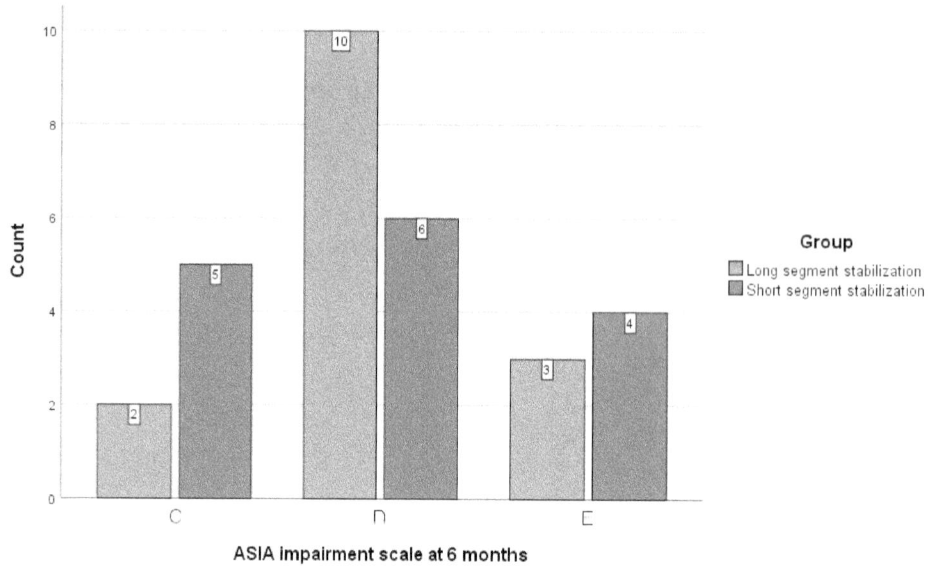
10
8
6
4
2
0
Count
10
6
2
5
3
4
C
D
E
ASIA impairment scale at 6 months
Group
Long segment stabilization
Short segment stabilization

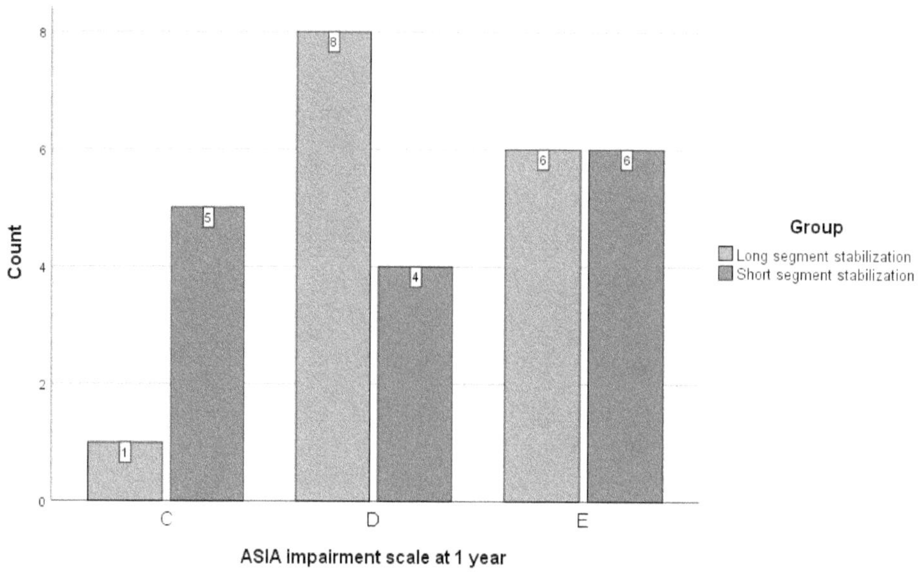
8
6
4
2
0
Count
8
6
6
5
4
1
C
D
E
ASIA impairment scale at 1 year
Group
Long segment stabilization
Short segment stabilization

Table 15: Longitudinal comparison of ASIA impairment scale in Long segment group (n=15)

		ASIA impairment scale at				p value
		Preoperative	3 months	6 months	1 year	
Long segment stabilization, n(%)	C	8 (53.3%)	5 (33.3%)	2 (13.3%)	1 (6.7%)	0.046#
	D	6 (40.0%)	8 (53.3%)	10 (66.7%)	8 (53.3%)	
	E	1 (6.7%)	2 (13.3%)	3 (20.0%)	6 (40.0%)	

p value by chi-square test; #-p<0.05

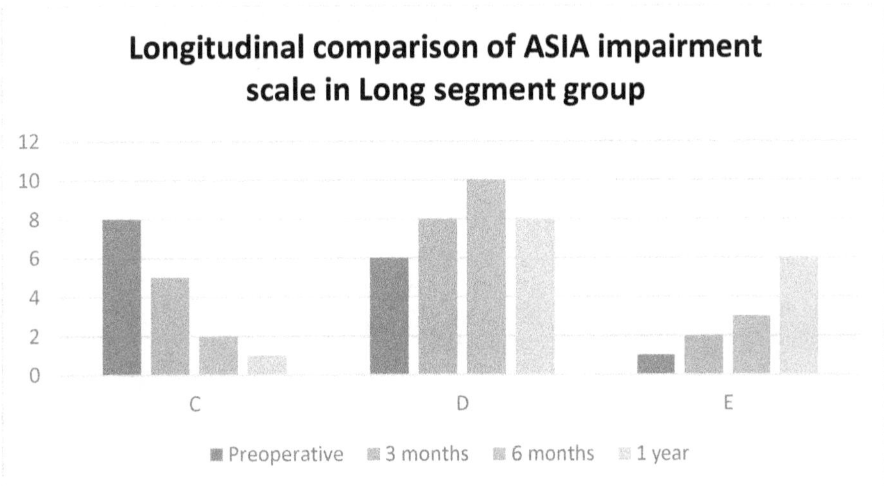

Figure 24: Graphical representation of the Longitudinal comparison of ASIA impairment scale in Long segment group (n=15)

The longitudinal comparison of the ASIA impairment scale at different timelines in the long segment group reports a shift from majority in the class C in the preoperative period to class D after one year. The comparison was statistically significant ($p<0.05$) inferring a significant trend in the ASIA impairment scale.

Table 16: Longitudinal comparison of ASIA impairment scale in short segment group (n=15)

		ASIA impairment scale at				p value
		Preoperative	3 months	6 months	1 year	
Short segment stabilization, n(%)	C	6 (40.0%)	6 (40.0%)	5 (33.3%)	5 (33.3%)	0.248
	D	8 (53.3%)	8 (53.3%)	6 (40.0%)	4 (26.7%)	
	E	1 (6.7%)	1 (6.7%)	4 (26.7%)	6 (40.0%)	

** p value by chi-square test*

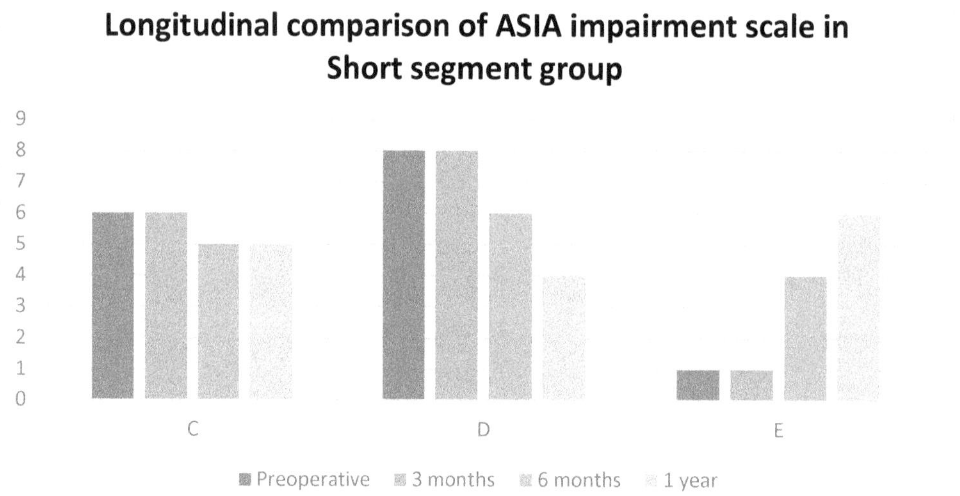

Figure 25: Graphical representation of the Longitudinal comparison of ASIA impairment scale in short segment group (n=15)

The longitudinal comparison of the ASIA impairment scale at different timelines in short segment group also reports a shift for a majority of patients in pre-operative period from class C to class D after one year. The comparison was statistically insignificant inferring no significant trend in the ASIA impairment scale.

Table 17: Beck's index comparison at different timelines among the intervention groups (n=30)

	Group, mean (SD)		p value*
	Long segment stabilization (n=15)	Short segment stabilization (n=15)	
Beck's Index-preoperative	0.6 (.07)	0.6 (.05)	0.541
Beck's index- 3 months	0.77 (.048)	0.77 (.050)	0.828
Beck's index- 6 months	0.8 (.05)	0.8 (.05)	0.886
Beck's index- 1 year	0.77 (.047)	0.76 (.049)	0.740

**p value by independent t test*

The above table shows the comparison of the Beck's index in the preoperative period and at 3, 6 and 12 months. The Beck's Index score was 0.6 (.07), 0.77 (.048), 0.8 (.05) and 0.77 (.047) in the long segment group. Similarly, Beck's Index score was 0.6 (.05), 0.77 (.050), 0.8 (.05) and 0.76 (.049) in the short segment group. The difference between the two groups was statistically insignificant at all timelines inferring equal distribution in both groups.

Table 18: Longitudinal comparison of the Beck's scores for each group (n=30)

Group	Timelines, mean (SD)				p value*
	Beck's Index-preoperative	Beck's index- 3 months	Beck's index- 6 months	Beck's index- 1 year	
Long segment stabilization (n=15)	0.6 (.07)	0.77 (.048)	0.8 (.05)	0.77 (.047)	<0.001#
Short segment stabilization (n=15)	0.6 (.05)	0.77 (.050)	0.8 (.05)	0.76 (.049)	<0.001#

**p value by RMNOVA; #-p<0.05*

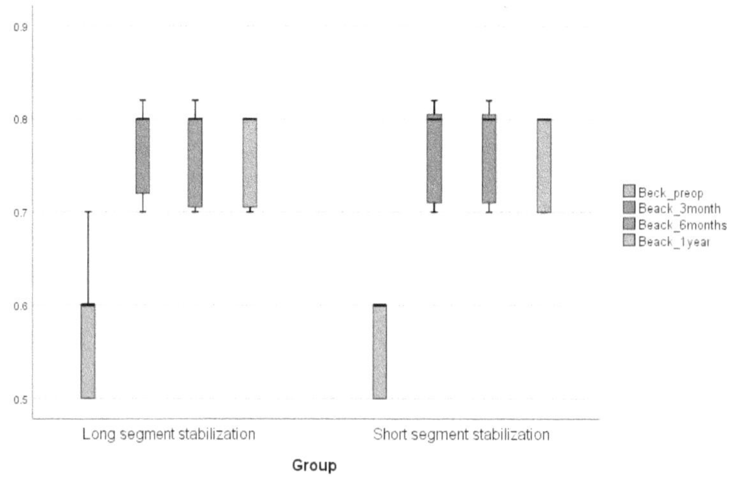

Figure 26: Graphical representation of the Longitudinal comparison of the Beck's scores for each group (n=30)

The longitudinal comparison of the becks index score at different timelines in both the long and short segment group shows a significant increase in the score (p<0.05) from the preoperative period to one year after surgery.

Table 19: Cobbs' angle comparison at different timelines among the groups (n=30)

	Group, mean(SD)		p value*
	Long segment stabilization (n=15)	Short segment stabilization (n=15)	
Cobbs angle- preoperative	16.1 (3.09)	17.2 (2.70)	0.323
Cobbs angle- 3 months	4.5 (2.50)	5.1 (2.43)	0.559
Cobbs angle- 6 months	4.6 (2.56)	5.1 (2.43)	0.615
Cobbs angle- 1 year	4.8 (2.46)	5.6 (2.72)	0.403

**p value by independent t test*

The above table shows the comparison of the Cobb's angle in the preoperative period and at 3, 6 and 12 months. The Cobb's angle was 16.1 (3.09), 4.5 (2.50), 4.6 (2.56), and 4.8 (2.46) in the long segment group. Similarly, Cobb's angle was 17.2 (2.70), 5.1 (2.43), 5.1 (2.43) and 5.6 (2.72) in the short segment group. The difference between the two groups was statistically insignificant at all timelines inferring equal distribution in both groups.

Table 20: Longitudinal comparison of the Cobb's angle for each group (n=30)

Group	Timelines, mean (SD)				p value*
	Cobbs angle- preoperative	Cobbs angle- 3 months	Cobbs angle- 6 months	Cobbs angle- 1 year	
Long segment stabilization (n=15)	16.1 (3.09)	4.5 (2.50)	4.6 (2.56)	4.8 (2.46)	<0.001#
Short segment stabilization (n=15)	17.2 (2.70)	5.1 (2.43)	5.1 (2.43)	5.6 (2.72)	<0.001#

**p value by RMNOVA; #-p<0.05*

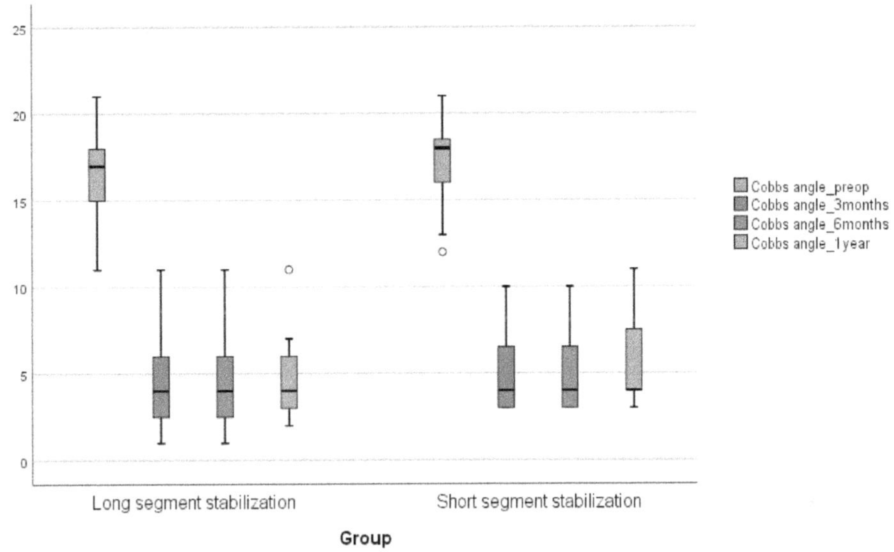

Figure 27: Graphical representation of the Longitudinal comparison of the Cobb's angle for each group (n=30)

The longitudinal comparison of the Cobb's angle at different timelines in both the long and short segment group shows a significant decrease in the Cobb's angle ($p<0.05$) from the preoperative period to one year after surgery.

Table 21: Comparison of Denis pain and work scale among the groups (n=30)

		Group, n(%)		p value*
		Long segment stabilization (n=15)	Short segment stabilization (n=15)	
Dennis pain scale	P2	6 (40.0%)	5 (33.3%)	0.096
	P3	4 (26.7%)	9 (60.0%)	
	P4	5 (33.3%)	1 (6.7%)	
Dennis work scale	W2	1 (6.7%)	1 (6.7%)	0.863
	W3	7 (46.7%)	5 (33.3%)	
	W4	6 (40.0%)	7 (46.7%)	
	W5	1 (6.7%)	2 (13.3%)	

* *p value by chi-square test*

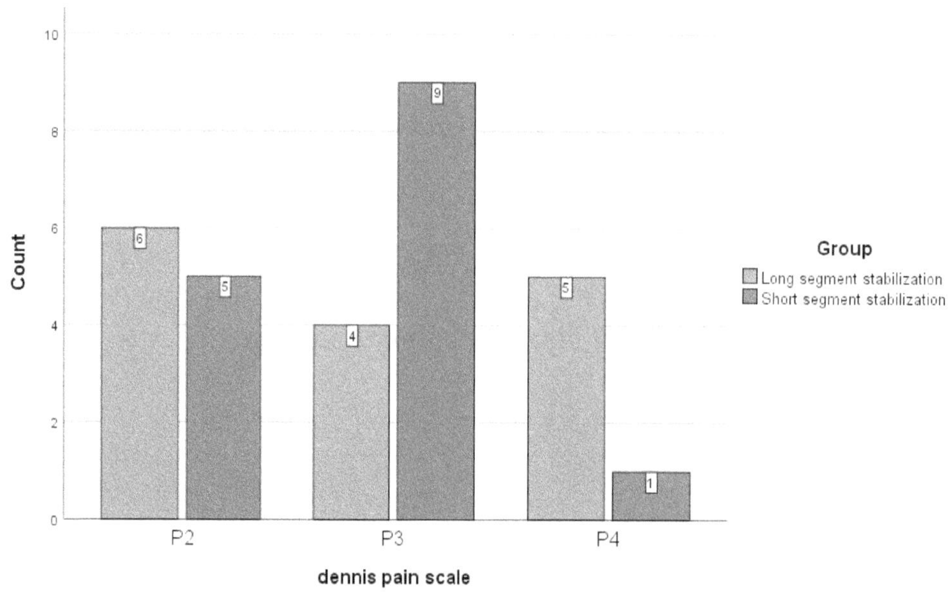

Figure 28: Graphical Comparison of Dennis pain scale among the groups (n=30)

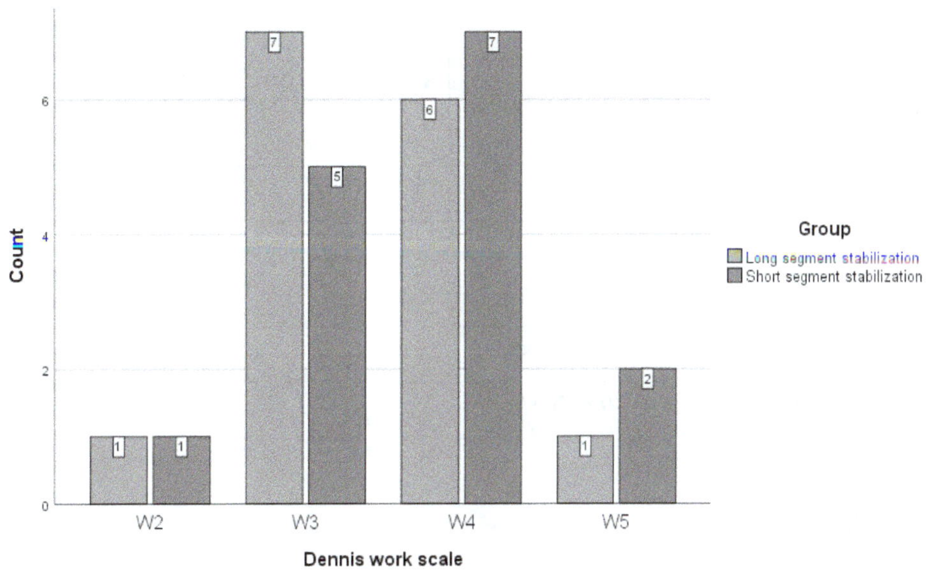

Figure 29: Graphical Comparison of Dennis pain scale among the groups (n=30)

Majority of the patients had P2 (40%) in the Denis pain scale and W3 (46.7%) in the Denis work scale in the long segment group while most of the patients had P3 (60%) in the Denis pain scale and W4 (46.7%) in the Denis work scale in the short segment group.

Table 22: Comparison of Complications among the intervention groups (n=30)

		Group, n(%)		p value*
		Long segment stabilization (n=15)	Short segment stabilization (n=15)	
Complications	Nil	13 (86.7%)	8 (53.3%)	0.136
	Urinary tract infections	1 (6.7%)	4 (26.7%)	
	Bed sores	1 (6.7%)	3 (20%)	

** p value by chi-square test*

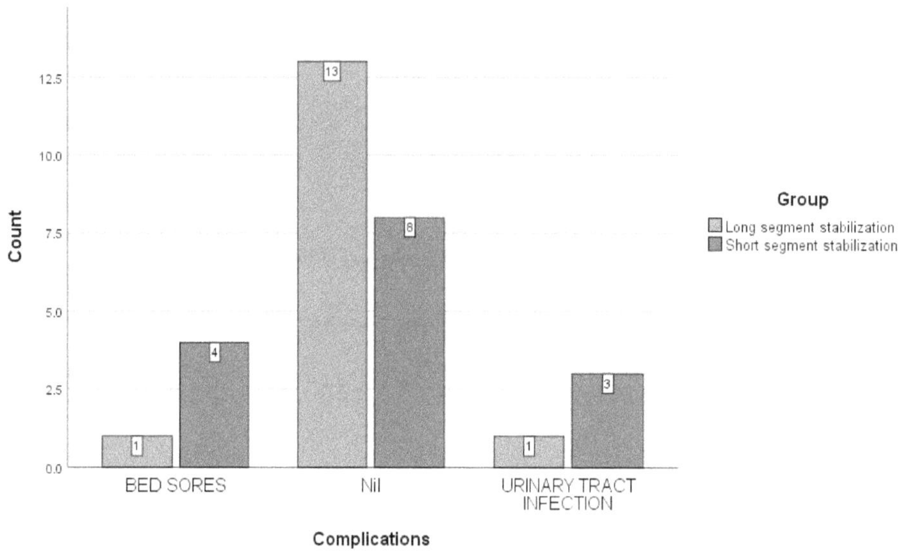

Figure 30: Graphical representation of the Comparison of Complications among the intervention groups (n=30)

In the long segment group, only one patient each had urinary tract infection and bed sore while for the short segment group it was four and three patients respectively who had urinary tract infection and bed sores.

Table 23: day of mobilization

Day of mobilization	Short segment	Long segment	P-value
1-2 days	9	8	0.124
2-12 days	3	7	
>12 days	3	1	

The day of mobilization was documented for all the patients. Almost all of them were mobilized within 12 days postoperatively, except for four patients [three being in the short segment group and 1 being in the long segment group].

Table 24: Crosslink augmentation

	Short segment	Long segment	P-value
With cross-links	3	2	0.309
Without cross-links	12	13	

In the short segment group 3 patients had their connecting rods fixed with cross-links, while 2 patients in the long segment group had their constructs augmented by cross-links.The rest of the patients in both groups did not have cross-link augmentation.

CASE 1

Figure 31

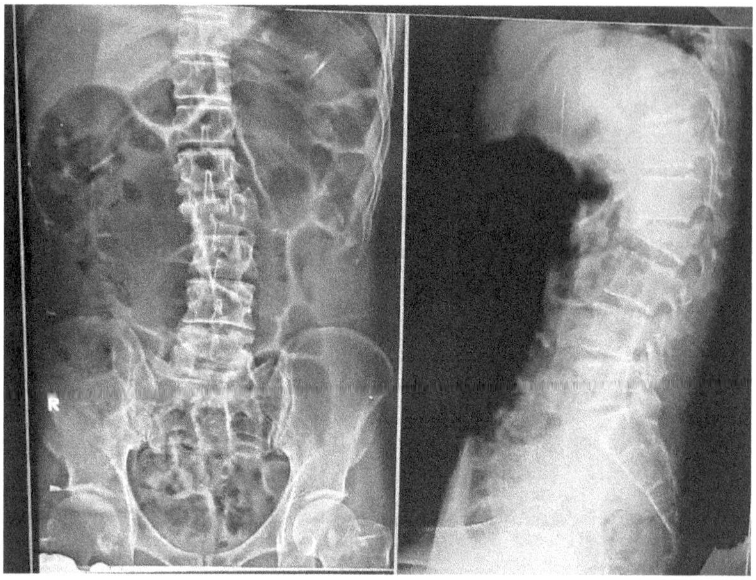

pre-op xrays

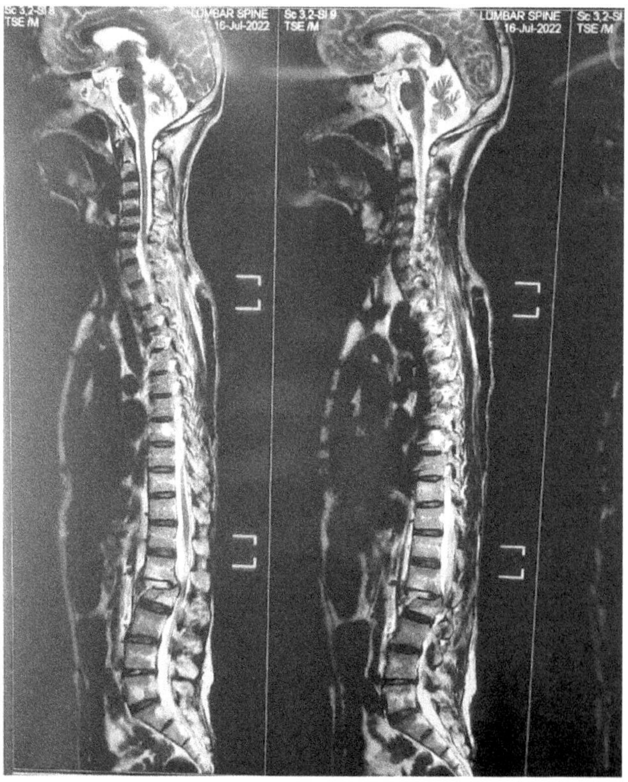

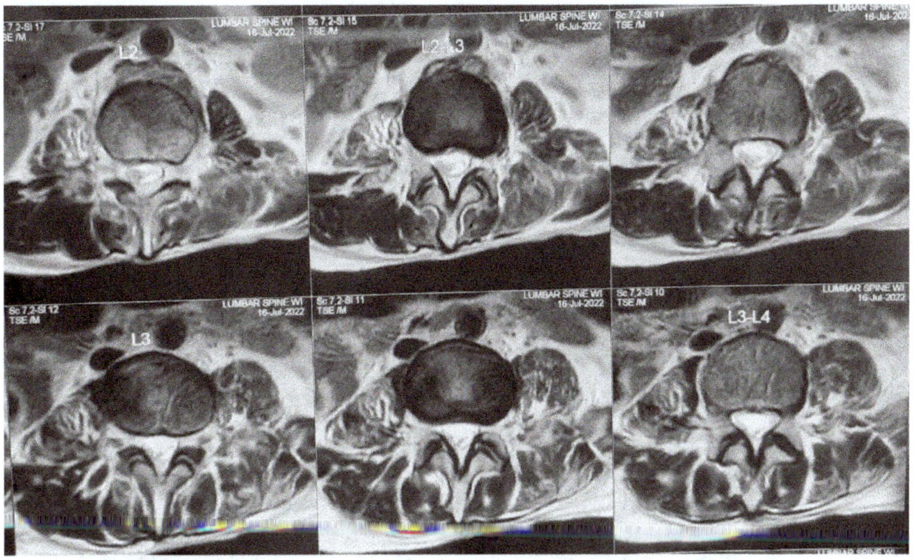
L2
L2-L3
L3
L3-L4

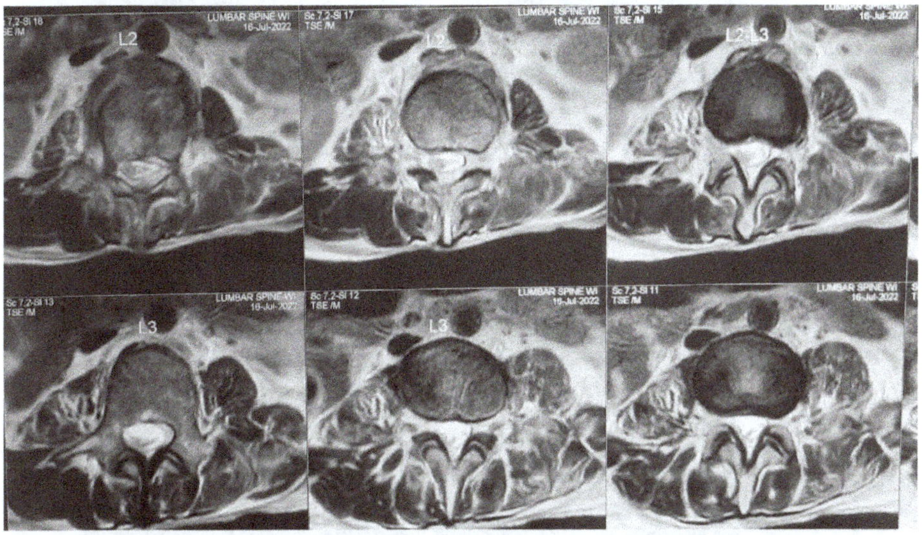
L2
L2
L2-L3
L3
L3

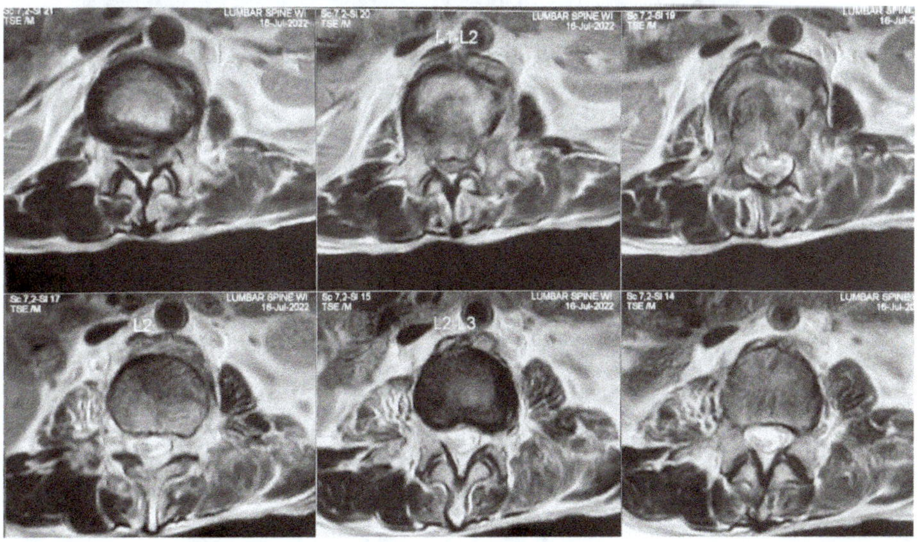
L1-L2
L2
L2-L3

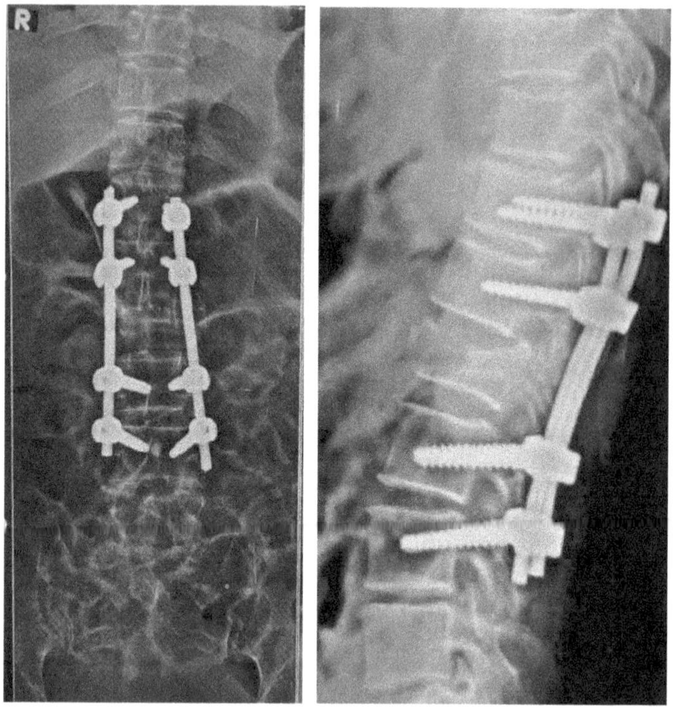

Immediate postop

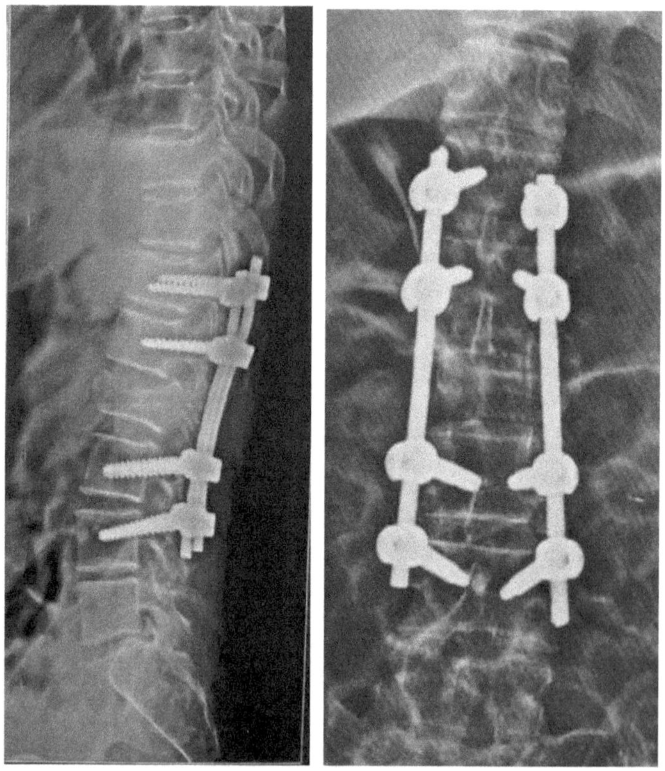

3 months follow-up

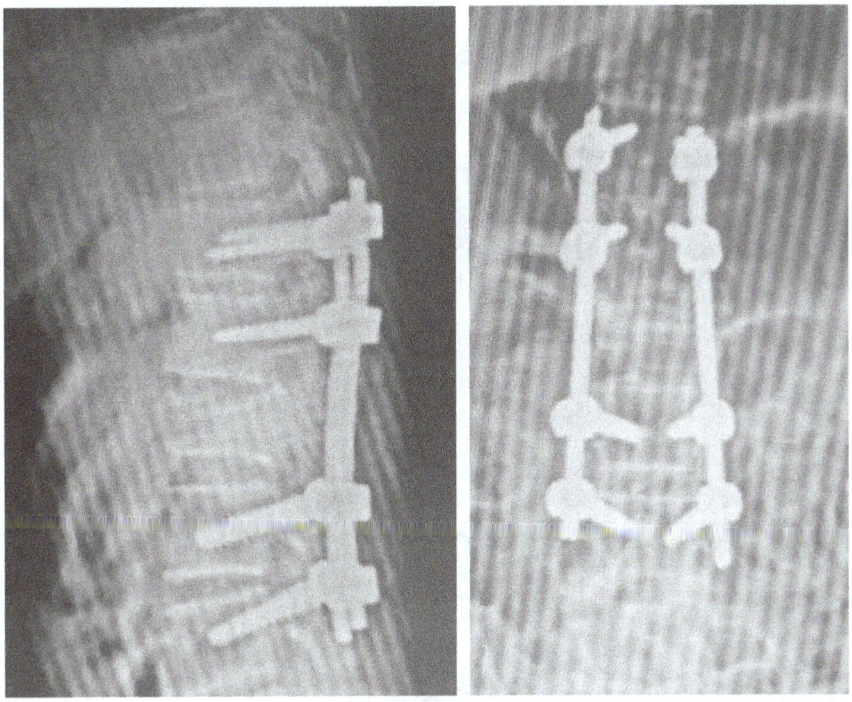

6 months follow-up

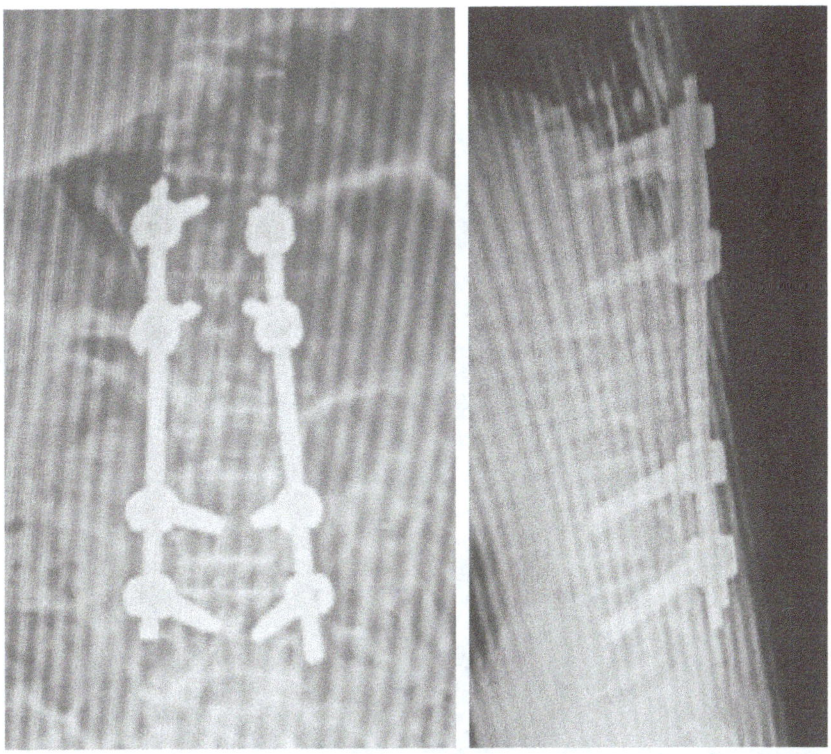

12 months follow-up

CASE 2

Figure 32

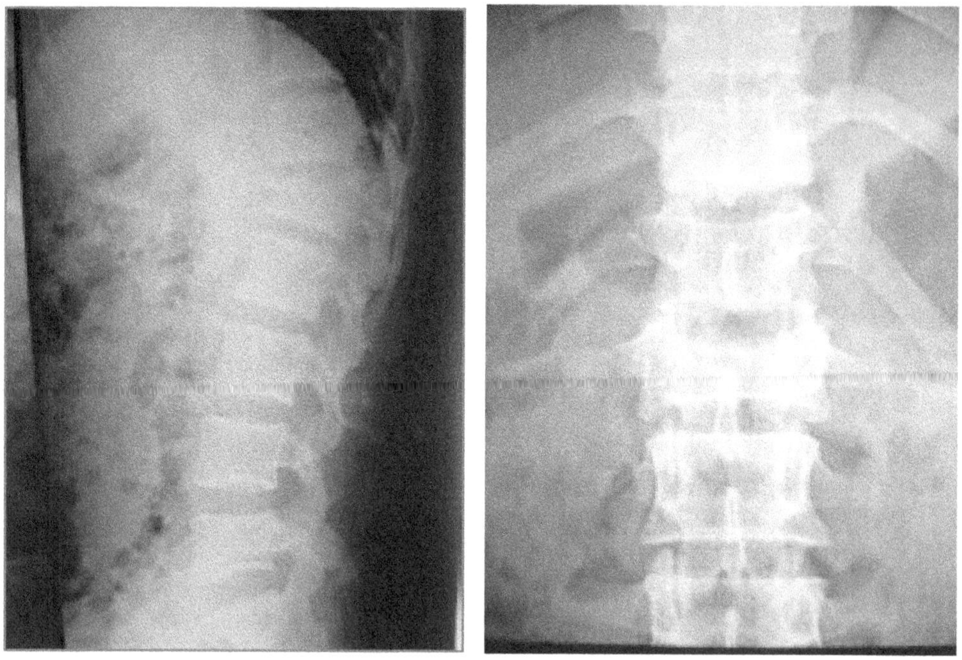

Pre-op

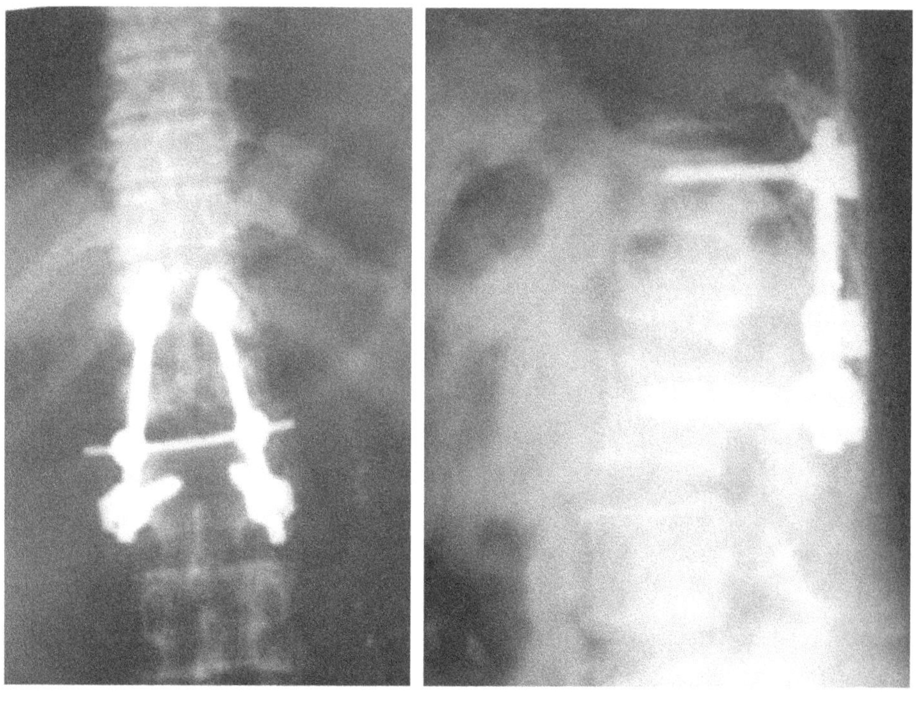

Immediate postop

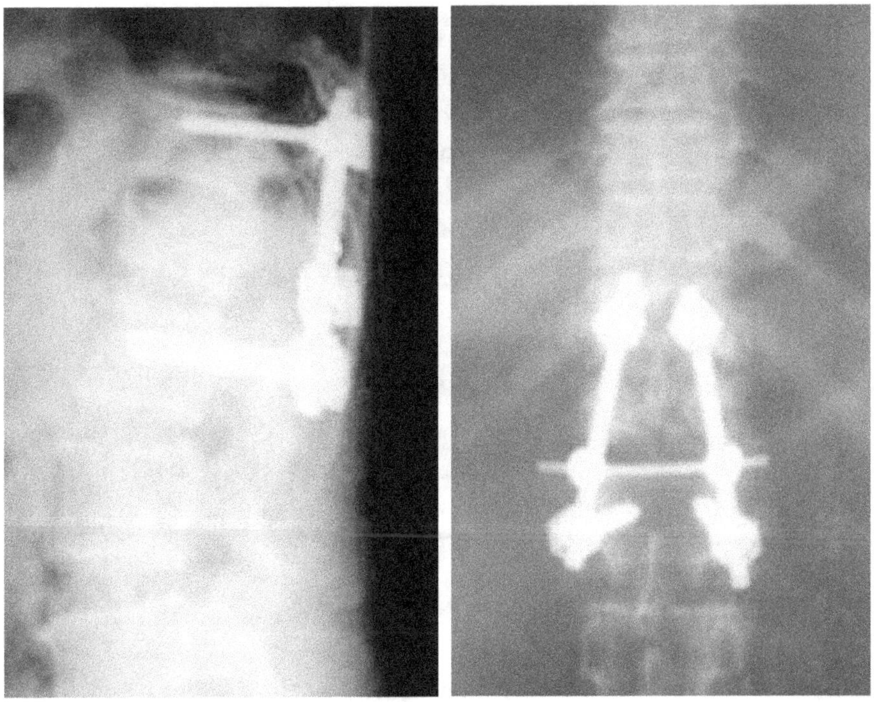

3 months follow-up

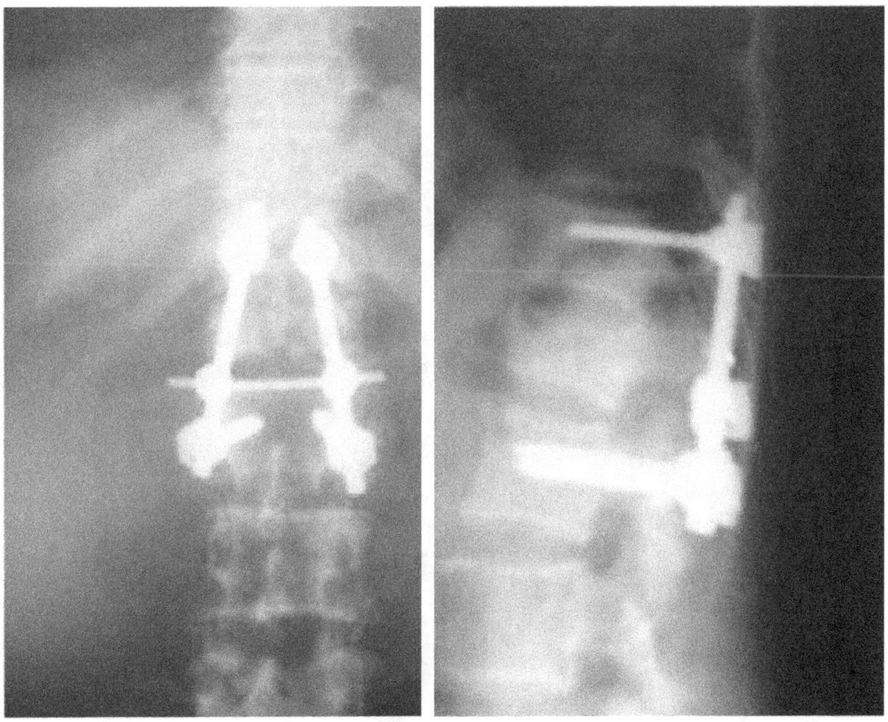

6 months follow-up

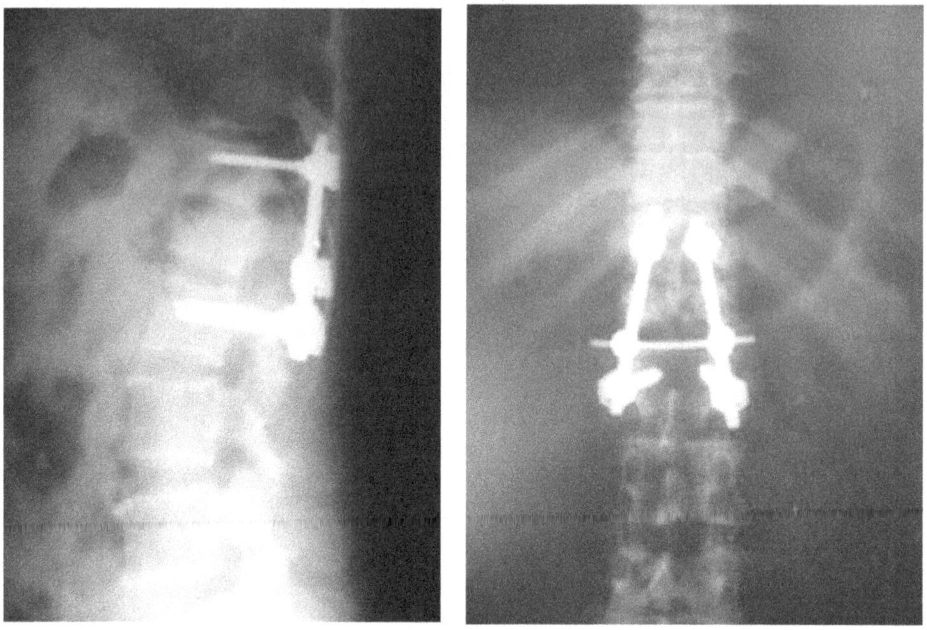

1 year follow-up

CASE 3

Figure 33

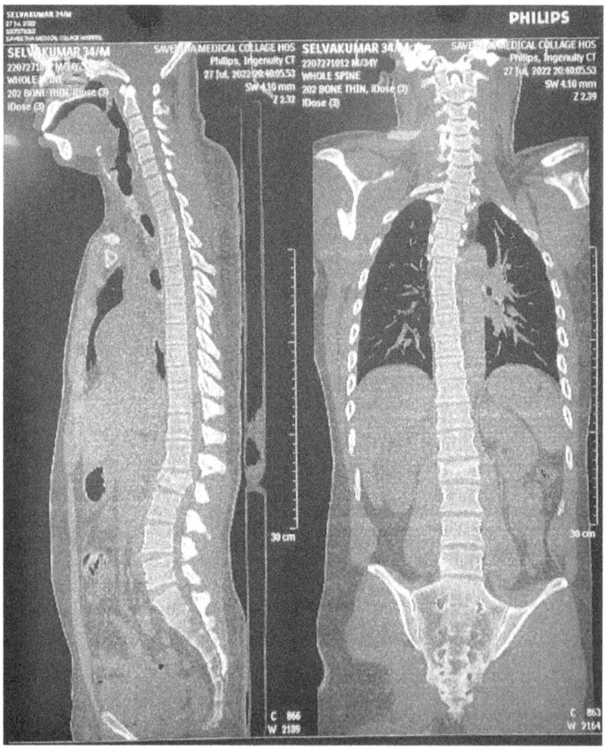

Pre-op

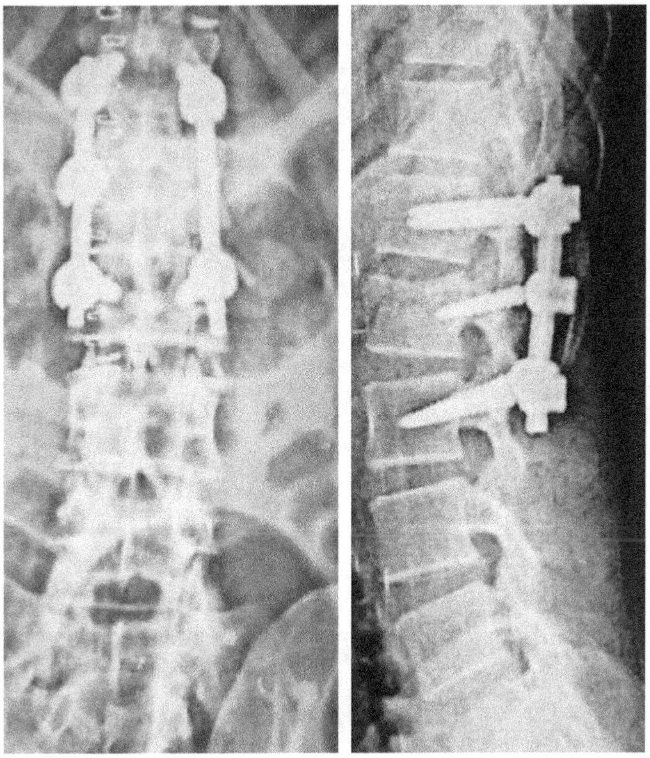

Immediate postop

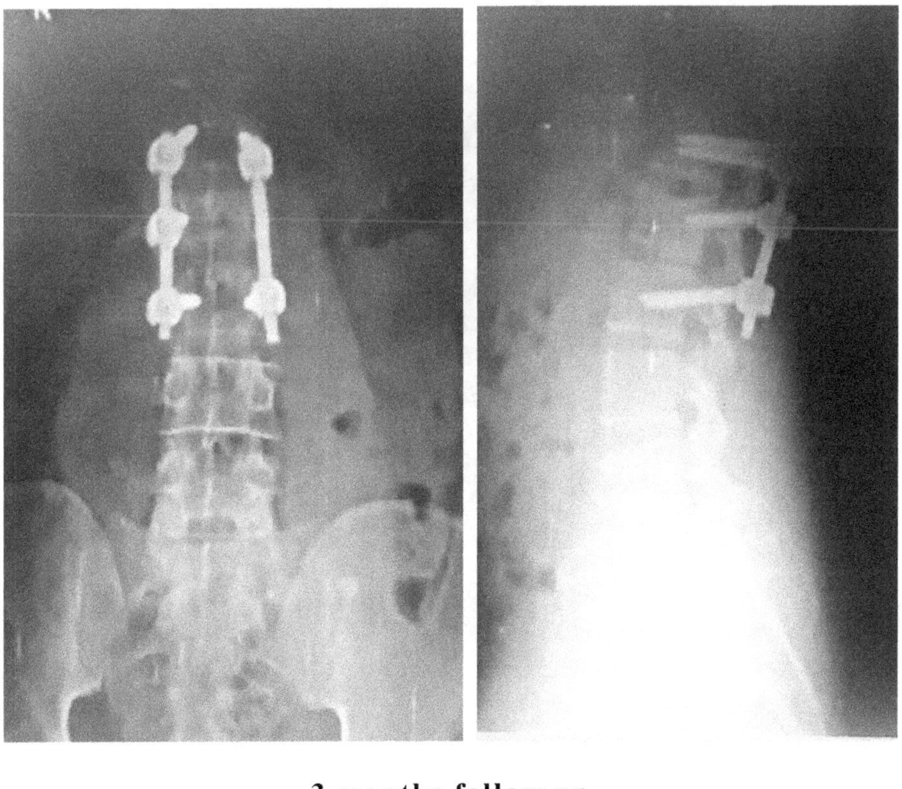

3 months follow up

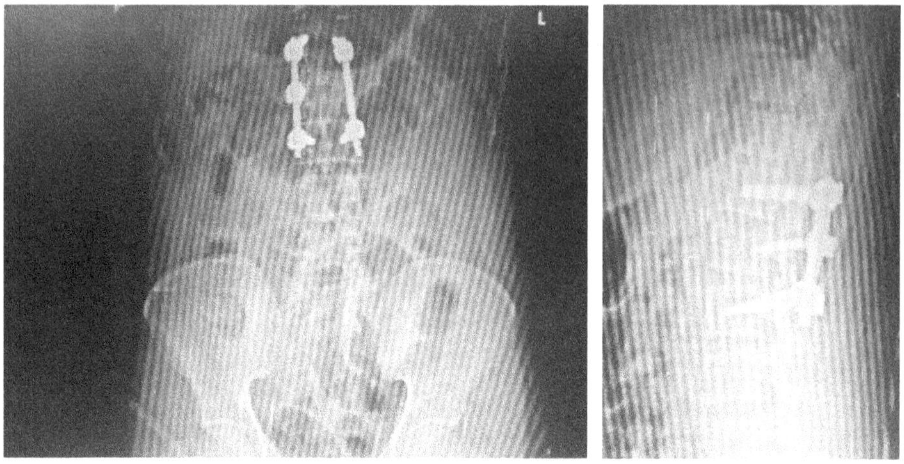

6 months follow-up

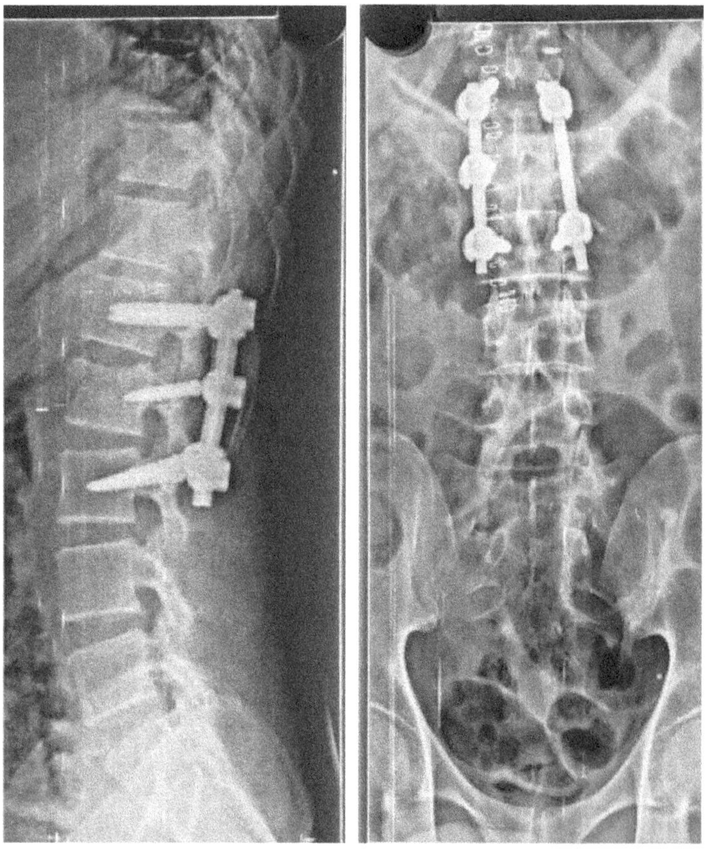

1 year follow-up

DISCUSSION

A dorsolateral spinal injury with neurological dysfunction is a orthopedic emergency that increases morbidity and death significantly. Previously used implants such as luque rods and sublaminar wires were associated with higher levels of implant failure and lesser union rates. Boucher [76] first described posterior transpedicular screw fixation in 1959. Transpedicular screws were found to be more sturdy and architecturally stable till bony fusion[77]. The most popular and straightforward procedure, short-segment posterior fixation, has the advantage of adding fewer motion segments into the fusion [71, 78, 79, 80]. In the studies with long-term follow-up, of the literature revealed that Short-segment posterior fixation alone was associated with a 9%–54% incidence of implant failure and re-kyphosis, and that 50% of the patients who experienced implant failure experienced moderate-to-severe pain [74, 78].

We had encountered, in our study, complications such as bed sores and urinary tract infections. In the long segment group, one patient each had bed sore and UTI. Meanwhile, in the short segment group, 4 patients had UTI and 3 patients had bed sores. The patients who had contracted UTIs had a urine sample collected and sent for culture and sensitivity, based on which they were started on appropriate antibiotics. The patients with bed sores had their positions changed frequently. They were provided with air/water mattresses to relieve pressure. Dressings were done. All the complications were resolved in a maximum of thiry two days.

Our studies showed no statistically significant differences in the postoperative period for the ASIA scores inferring equal distribution in both groups. Several studies revealed that short segment pedicle instrumentation has a significant failure rate even though it was presumed to be a superior technique. Three types of hardware failure with this fixation were noted in their report on the early failure of short segment pedicle fixation for thoracolumbar burst fracture: progressive kyphosis secondary to screw bending, kyphosis secondary to osseous collapse or vertebral translation without hardware bending, and segmental kyphosis following the breakage of a caudad screw in the lumbar construct. They also observed that pre-stressing the screws when the rods were shaped in situ and untreated anterior instability both contributed to a high failure rate [73, 74, 82, 83]. According to Altay et al. [72], lengthening the level arm of the construct with four pairs of screws (two above and two below) would likely increase stability while also effectively reducing kyphotic deformity. For some fractures in the thoracolumbar junction, short segment posterior fixation alone can produce positive clinical and radiological outcomes. In order to avoid implant failure, corrective loss, and to meet the requirement for anterior column support, it is crucial to identify fractures that can be treated with short-segment posterior fixation [72]. Our research revealed that the operating time and blood loss was more in the long segment fixation group. In their study comparing short segment with long segment fixation in thoracolumbar burst fractures, Tezeren and Kuru [71] showed that long segment instrumentation is a successful treatment option. Instrumentation for short segment fixation of the pedicle has a high failure rate. However, lengthy segment instrumentation lengthened the duration of surgical procedure and subsequently increased blood loss, which was consistent with the findings of our investigation. De Peretti et al. [81] claimed that fixation by screw and hook structures, holding the two vertebrae above the lesion and screws and hooks holding the first vertebra below the lesion, was a successful method to stabilise thoracolumbar junction burst fractures. At the thoracolumbar junction, where compressive pressures act more anteriorly, segmental transpedicular fixation two levels above the kyphosis should be employed, according to Carl et al. To avoid progressive kyphosis and equipment failure, they preferred to position the pedicle screw two levels above the fracture site. On the other hand, one level fixation adjacent to the fracture site was preferred to preserve the mobility segment in the lumbar level as much as possible. Short segment pedicle screw fixation in thoracolumbar burst fractures was successful, according to Butt et al. [84], although the 40% hardware failure rate they documented (20 out of 50 patients) is concerning. In an unstable calf spine model, Gurr and

McAfee [85] discovered that two levels above and below the wounded level gave higher stiffness than the intact spine. Tarek Et al. found that in all stress types, hybrid and extended posterior fixation considerably increased stability relative to intact and wounded situations. Furthermore, compared to a brief posterior fixation, both fixations offered stronger stability. In light of this, lengthy posterior fixation seems to be a successful treatment for thoracolumbar burst fractures [85].

In their biomechanical comparison of long segment versus short segment fixation, Peters et al. [87] discovered that long posterior fixation was the stiffest under all loading situations, both with and without anterior column augmentation. It has been demonstrated that using screws two above and two below not only increases stability but also effectively reduces kyphotic deformity [88]. The more lordotic middle and lower lumbar spine, where the compressive pressures operate more posteriorly, has been demonstrated to have fractures that can be successfully treated with short posterior fixation alone to produce positive clinical and radiological outcomes. One level above and one level below the fracture in the lumbar region formed a rigid construct with no corrective loss, according to Katonis et al. [70].

The least amount of disruption to the patients' life is achieved with surgical care of thoracolumbar fractures. According to the patients, two important outcomes for surgical treatment of thoracolumbar burst fractures are pain alleviation and radiological rectification [88]. Patients who underwent posterior short segment fixation or posterior long segment fixation were evaluated using the Denis Pain and Work Scale over the follow-up periods [69, 75]. Even though the short segment group's pain and return to work scores improved less, the differences were not statistically significant, therefore both results on the denis scale could be compared. According to our research, kyphosis progresses in both groups even though there was no difference between the short segment fixation and long segment fixation groups at the time of the last follow-up. Much of this progression also seems to occur in the first few days following injury, according to Xu et al. [89].

We acknowledge some limitations in our study. The sample size was 30 patients with 15 being in each group. A larger sample size would have been more helpful to analyse the differences. The follow-up period was for a duration of 1 year minimum. Although we did not encounter significant differences apart from operative time and blood loss, there could be statistically significant differences which can only be noted if there was a longer follow-up. It was not a randomised controlled trial. As such, this study hopefully will facilitate greater evidence-based practice and quality management. `

CONCLUSION

Through this study ,the results measured for the radiological and clinical outcome were not statistically significant inferring equal distribution for both procedures leading to comparable results. Although it should be noted that long segment posterior stabilization did involve a longer operative time and is associated with higher levels of blood loss, both of which were statistically significant. Future studies with larger sample sizes and having a randomized control trial would be needed along with a longer follow-up period.

In conclusion, the results of this study showed that in the surgical management of thoracolumbar burst fractures through posterior approach, the short segment pedicle screws fixation was not significantly different to the long segment pedicle screws fixation in terms of improvement of back pain and return to work and also in correction of kyphotic deformity.

BIBLIOGRAPHY

1. Rajasekaran S, Kanna RM, Shetty AP. Management of thoracolumbar spine trauma: An overview. Indian J Orthop. 2015 Jan-Feb;49(1):72-82. doi: 10.4103/0019-5413.143914. PMID: 25593358; PMCID: PMC4292328
2. Bucholz RW, Heckman JD, Tornetta P, McQueen MM, Ricci WM. Rockwood and Green's fractures in adults. In: Rockwood and Green's fractures in adults. 2010. p. 1275–1275.
3. Charles YP, Steib JP. Management of thoracolumbar spine fractures with neurologic disorder. Orthop Traumatol Surg Res. 2015 Feb; 101(1 Suppl): S31-40. doi: 10.1016/j.otsr.2014.06.024. Epub 2015 Jan 7. PMID: 25577599
4. Glaser SE, Falco F. Paraplegia following a thoracolumbar transforaminal epidural steroid injection. Pain Physician. 2005 Jul;8(3):309-14. PMID: 16850088
5. Sehgal N, Dunbar EE, Shah RV, Colson J. Systematic review of diagnostic utility of facet (zygapophysial) joint injections in chronic spinal pain: an update. Pain Physician. 2007 Jan;10(1):213-28. PMID: 17256031

6. Dimar JR, Fisher C, Vaccaro AR, Okonkwo DO, Dvorak M, Fehlings M, Rampersaud R, Carreon LY. Predictors of complications after spinal stabilization of thoracolumbar spine injuries. J Trauma. 2010 Dec;69(6):1497-500. doi: 10.1097/TA.0b013e3181cc853b. PMID: 20404758.
7. Chapman MW, James MA. Chapman's Comprehensive Orthopaedic Surgery: Five Volume Set. Jaypee Brothers Medical Publishers; 2019.
8. Denis F. Spinal instability as defined by the three-column spine concept in acute spinal trauma. Clin Orthop Relat Res. 1984 Oct;(189):65-76. PMID: 6478705.
9. Dai LY, Jiang SD, Wang XY, Jiang LS. A review of the management of thoracolumbar burst fractures. Surg Neurol. 2007 Mar;67(3):221-31; discussion 231. doi: 10.1016/j.surneu.2006.08.081. PMID: 17320622.
10. Sasso RC, Renkens K, Hanson D, Reilly T, McGuire RA Jr, Best NM. Unstable thoracolumbar burst fractures: anterior-only versus short-segment posterior fixation. J Spinal Disord Tech. 2006 Jun;19(4):242-8. doi: 10.1097/01.bsd.0000211298.59884.24. PMID: 16778657.
11. Aly TA. Short Segment versus Long Segment Pedicle Screws Fixation in Management of Thoracolumbar Burst Fractures: Meta-Analysis. Asian Spine J. 2017 Feb;11(1):150-160. doi: 10.4184/asj.2017.11.1.150. Epub 2017 Feb 17. PMID: 28243383; PMCID: PMC5326724.

12. Korovessis P, Baikousis A, Zacharatos S, Petsinis G, Koureas G, Iliopoulos P. Combined anterior plus posterior stabilization versus posterior short-segment instrumentation and fusion for mid-lumbar (L2-L4) burst fractures. Spine (Phila Pa 1976). 2006 Apr 15;31(8):859-68. doi: 10.1097/01.brs.0000209251.65417.16. PMID: 16622372.

13. Wahba GM, Bhatia N, Bui CN, Lee KH, Lee TQ. Biomechanical evaluation of short-segment posterior instrumentation with and without crosslinks in a human cadaveric unstable thoracolumbar burst fracture model. Spine (Phila Pa 1976). 2010 Feb 1;35(3):278-85. doi: 10.1097/BRS.0b013e3181bda4e6. PMID: 20075769.

14. Dobran M, Nasi D, Brunozzi D, di Somma L, Gladi M, Iacoangeli M, Scerrati M. Treatment of unstable thoracolumbar junction fractures: short-segment pedicle fixation with inclusion of the fracture level versus long-segment instrumentation. Acta Neurochir (Wien). 2016 Oct;158(10):1883-9. doi: 10.1007/s00701-016-2907-0. Epub 2016 Aug 19. PMID: 27541493.

15. Waqar M, Van-Popta D, Barone DG, Bhojak M, Pillay R, Sarsam Z. Short versus long-segment posterior fixation in the treatment of thoracolumbar junction fractures: a comparison of outcomes. Br J Neurosurg. 2017 Feb;31(1):54-57. doi: 10.1080/02688697.2016.1206185. Epub 2016 Jul 8. PMID: 27387358.

16. Canale & Beaty: Campbell's Operative Orthopaedics, 11th ed. P 1729

17. Mahadevan V. Anatomy of the vertebral column. Surgery (Oxford). 2018 Jul 1;36(7):327-32.

18. Neeley OJ, El Ahmadieh TY, Kafka B, Bagley CA. Anatomy of the spinal cord. InMyelopathy 2022 (pp. 1-16). Springer, Cham.

19. Ganapathy MK, Reddy V, Tadi P. Neuroanatomy, spinal cord morphology. InStatPearls [Internet] 2021 Oct 30. StatPearls Publishing.

20. Hernandez-Gerez E, Fleming IN, Parson SH. A role for spinal cord hypoxia in neurodegeneration. Cell Death Dis. 2019 Nov 13;10(11):861. doi: 10.1038/s41419-019-2104-1. PMID: 31723121; PMCID: PMC6853899.

21. McLain RF. The biomechanics of long versus short fixation for thoracolumbar spine fractures. Spine (Phila Pa 1976). 2006 May 15;31(11 Suppl):S70-9; discussion S104. doi: 10.1097/01.brs.0000218221.47230.dd. PMID: 16685240.

22. Panjabi MM, Brand RA Jr, White AA 3rd. Mechanical properties of the human thoracic spine as shown by three-dimensional load-displacement curves. J Bone Joint Surg Am. 1976 Jul;58(5):642-52. PMID: 932062.

23. Panjabi MM, Hausfeld JN, White AA 3rd. A biomechanical study of the ligamentous stability of the thoracic spine in man. Acta Orthop Scand. 1981 Jun;52(3):315-26. doi: 10.3109/17453678109050109. PMID: 7282325.

24. Langrana NA, Harten RD RD, Lin DC, Reiter MF, Lee CK. Acute thoracolumbar burst fractures: a new view of loading mechanisms. Spine (Phila Pa 1976). 2002 Mar 1;27(5):498-508. doi: 10.1097/00007632-200203010-00010. PMID: 11880835..

25. Denis F. The three column spine and its significance in the classification of acute thoracolumbar spinal injuries. Spine (Phila Pa 1976). 1983 Nov-Dec;8(8):817-31. doi: 10.1097/00007632-198311000-00003. PMID: 6670016..

26. Su Q, Li C, Li Y, Zhou Z, Zhang S, Guo S, Feng X, Yan M, Zhang Y, Zhang J, Pan J, Cheng B, Tan J. Analysis and improvement of the three-column spinal theory. BMC Musculoskelet Disord. 2020 Aug 12;21(1):537. doi: 10.1186/s12891-020-03550-5. PMID: 32787828; PMCID: PMC7425572.

27. Diamante GA, Marinho PV, Zani CC, Arias MV. Ex-vivo evaluation of the three-column concept in canine thoracolumbar fractures. Arquivo Brasileiro de Medicina Veterinária e Zootecnia. 2020 Aug 14;72:1221-30.

28. Costachescu B, Popescu CE, Iliescu BF. Analysis of the Classification Systems for Thoracolumbar Fractures in Adults and Their Evolution and Impact on Clinical Management. J Clin Med. 2022 Apr 29;11(9):2498. doi: 10.3390/jcm11092498. PMID: 35566623; PMCID: PMC9100632.

29. Lee JY, Vaccaro AR, Lim MR, Oner FC, Hulbert RJ, Hedlund R, Fehlings MG, Arnold P, Harrop J, Bono CM, Anderson PA, Anderson DG, Harris MB, Brown AK, Stock GH, Baron EM. Thoracolumbar injury classification and severity score: a new paradigm for the treatment of thoracolumbar spine trauma. J Orthop Sci. 2005 Nov;10(6):671-5. doi: 10.1007/s00776-005-0956-y. PMID: 16307197; PMCID: PMC2779435.

30. Ergashov AR. Modern Clinical Analysis of Injuries of the Thoracolumbar Spine. International journal of health systems and medical sciences. 2022 Oct 21;1(4):59-63.

31. Arocho-Quiñones E, Soliman H, Kurpad S. A Comprehensive Classification of Thoracic and Lumbar Injuries. In50 Landmark Papers 2018 Jul 24 (pp. 77-81). CRC Press.

32. Filgueira ÉG, Imoto AM, da Silva HEC, Meves R. Thoracolumbar Burst Fracture: McCormack Load-sharing Classification: Systematic Review and Single-arm Meta-analysis. Spine (Phila Pa 1976). 2021 May 1;46(9):E542-E550. doi: 10.1097/BRS.0000000000003826. PMID: 33273433.

33. Vu C, Gendelberg D. Classifications in Brief: AO Thoracolumbar Classification System. Clin Orthop Relat Res. 2020 Feb;478(2): 434 - 440. doi: 10.1097/CORR.0000000000 001086. PMID: 32022766; PMCID: PMC7438146.

34. Curt A, Keck ME, Dietz V. Functional outcome following spinal cord injury: significance of motor-evoked potentials and ASIA scores. Arch Phys Med Rehabil. 1998 Jan;79(1):81-6. doi: 10.1016/s0003-9993(98)90213-1. PMID: 9440423..

35. Kirshblum SC, Burns SP, Biering-Sorensen F, Donovan W, Graves DE, Jha A, Johansen M, Jones L, Krassioukov A, Mulcahey MJ, Schmidt-Read M, Waring W. International standards for neurological classification of spinal cord injury (revised 2011). J Spinal Cord Med. 2011 Nov;34(6):535-46. doi: 10.1179/204577211X13207446293695. PMID: 22330108; PMCID: PMC3232636..

36. Capaul M, Zollinger H, Satz N, Dietz V, Lehmann D, Schurch B. Analyses of 94 consecutive spinal cord injury patients using ASIA definition and modified Frankel score classification. Paraplegia. 1994 Sep;32(9):583-7. doi: 10.1038/sc.1994.92. PMID: 7997337..

37. Roberts TT, Leonard GR, Cepela DJ. Classifications In Brief: American Spinal Injury Association (ASIA) Impairment Scale. Clin Orthop Relat Res. 2017 May;475(5):1499-1504. doi: 10.1007/s11999-016-5133-4. Epub 2016 Nov 4. PMID: 27815685; PMCID: PMC5384910.

38. Khorasanizadeh M, Yousefifard M, Eskian M, Lu Y, Chalangari M, Harrop JS, Jazayeri SB, Seyedpour S, Khodaei B, Hosseini M, Rahimi-Movaghar V. Neurological recovery following traumatic spinal cord injury: a systematic review and meta-analysis. J Neurosurg Spine. 2019 Feb 15:1-17. doi: 10.3171/2018.10.SPINE18802. Epub ahead of print. PMID: 30771786..

39. Waters RL, Adkins RH, Yakura JS. Definition of complete spinal cord injury. Paraplegia. 1991 Nov;29(9):573-81. doi: 10.1038/sc.1991.85. PMID: 1787981.

40. McCourt TR. Spinal cord injuries. Trauma Nursing E-Book: From Resuscitation Through Rehabilitation. 2019 Aug 18:454.

41. Foo D, Subrahmanyan TS, Rossier AB. Post-traumatic acute anterior spinal cord syndrome. Paraplegia. 1981;19(4):201-5. doi: 10.1038/sc.1981.42. PMID: 7290729.

42. McKinley W, Hills A, Sima A. Posterior cord syndrome: Demographics and rehabilitation outcomes. J Spinal Cord Med. 2021 Mar;44(2):241 - 246. doi: 10.1080/10790268. 2019.1585135. Epub 2019 Apr 2. PMID: 30939076; PMCID: PMC7952058.

43. Roth EJ, Park T, Pang T, Yarkony GM, Lee MY. Traumatic cervical Brown-Sequard and Brown-Sequard-plus syndromes: the spectrum of presentations and outcomes. Paraplegia. 1991 Nov;29(9):582-9. doi: 10.1038/sc.1991.86. PMID: 1787982.

44. Rider LS, Marra EM. Cauda equina and conus medullaris syndromes. InStatPearls [Internet] 2021 Aug 11. StatPearls Publishing.

45. Priebe MM, Waring WP. The interobserver reliability of the revised American Spinal Injury Association standards for neurological classification of spinal injury patients. Am J Phys Med Rehabil. 1991 Oct;70(5):268-70. doi: 10.1097/00002060-199110000-00007. PMID: 1910652.

46. Medina GI, Nascimento FB, Rimkus CM, Zoppi Filho A, Cliquet A Jr. Clinical and radiographic evaluation of the shoulder of spinal cord injured patients undergoing rehabilitation program. Spinal Cord. 2011 Oct;49(10):1055-61. doi: 10.1038/sc.2011.64. Epub 2011 Jul 5. PMID: 21727901.

47. van Den Hauwe L, Sundgren PC, Flanders AE. Spinal Trauma and Spinal Cord Injury (SCI) 2020 Feb 15. In: Hodler J, Kubik-Huch RA, von Schulthess GK, editors. Diseases of the Brain, Head and Neck, Spine 2020–2023: Diagnostic Imaging [Internet]. Cham (CH): Springer; 2020. Chapter 19. Available from: https://www.ncbi.nlm.nih.gov/books/NBK554330/ doi: 10.1007/978-3-030-38490-6_19

48. Miyanji F, Furlan JC, Aarabi B, Arnold PM, Fehlings MG. Acute cervical traumatic spinal cord injury: MR imaging findings correlated with neurologic outcome--prospective study with 100 consecutive patients. Radiology. 2007 Jun;243(3):820-7. doi: 10.1148/radiol.2433060583. Epub 2007 Apr 12. PMID: 17431129..

49. Nori S, Watanabe K, Takeda K, Yamane J, Kono H, Yokogawa N, Sasagawa T, Ando K, Nakashima H, Segi N, Funayama T. Does surgery improve neurological outcomes in older individuals with cervical spinal cord injury without bone injury? A multicenter study. Spinal cord. 2022 Oct;60(10):895-902.

50. Vaccaro AR, Zeiller SC, Hulbert RJ, Anderson PA, Harris M, Hedlund R, Harrop J, Dvorak M, Wood K, Fehlings MG, Fisher C, Lehman RA Jr, Anderson DG, Bono CM, Kuklo T, Oner FC. The thoracolumbar injury severity score: a proposed treatment algorithm. J Spinal Disord Tech. 2005 Jun;18(3):209-15. PMID: 15905761.

51. Karaali E, Ciloglu O, Duramaz A, Kusvuran Ozkan A, Ekiz T. Management of thoracolumbar injury classification and severity score of 4 (TLICS=4) thoracolumbar vertebra fractures: Surgery versus conservative treatment. Ulus Travma Acil Cerrahi Derg. 2020 Sep;26(5):805-810. English. doi: 10.14744/tjtes.2020.30524. PMID: 32946086..

52. El Masri Y WS. Traumatic spinal injury and spinal cord injury: point for active physiological conservative management as compared to surgical management. Spinal Cord Ser Cases. 2018 Feb 22;4:14. doi: 10.1038/s41394-018-0045-z. PMID: 29479483; PMCID: PMC5823881.

53. Defino HLA, Costa HRT, Nunes AA, Nogueira Barbosa M, Romero V. Open versus minimally invasive percutaneous surgery for surgical treatment of thoracolumbar spine fractures- a multicenter randomized controlled trial: study protocol. BMC Musculoskelet Disord. 2019 Aug 31;20(1):397. doi: 10.1186/s12891-019-2763-1. PMID: 31472691; PMCID: PMC6717640..

54. Tan T, Rutges J, Marion T, Gonzalvo A, Mathew J, Fitzgerald M, Dvorak M, Schroeder G, Tee J. Anterior versus posterior approach in traumatic thoracolumbar burst fractures deemed for surgical management: Systematic review and meta-analysis. J Clin Neurosci. 2019 Dec;70:189-197. doi: 10.1016/j.jocn.2019.07.083. Epub 2019 Aug 17. PMID: 31431406..

55. Liang C, Liu G, Liang G, Zheng X, Yin D, Xiao D, Zeng S, Cai H, Chang Y. Healing pattern classification for thoracolumbar burst fractures after posterior short-segment fixation. BMC Musculoskelet Disord. 2020 Jun 12;21(1):373. doi: 10.1186/s12891-020-03386-z. PMID: 32532236; PMCID: PMC7291420.

56. Wang HW, Hu YC, Wu ZY, Wu HR, Ma JQ, Jian HQ, Ning SH, Xu WK. One Approach Anterior Decompression and Fixation with Posterior Unilateral Pedicle Screw Fixation for Thoracolumbar Osteoporotic Vertebral Compression Fractures. Orthop Surg. 2021 May;13(3):908-919. doi: 10.1111/os.12947. Epub 2021 Mar 30. PMID: 33787067; PMCID: PMC8126897..

57. Lin CL, Chou PH, Fang JJ, Huang KY, Lin RM. Short-segment decompression and fixation for thoracolumbar osteoporotic fractures with neurological deficits. J Int Med Res. 2018 Aug;46(8):3104 - 3113. doi: 10.1177/0300060 518772422. Epub 2018 Jun 8. PMID: 29882444; PMCID: PMC6134660.

58. Eichholz KM, Rabb CH, Anderson PA, Arnold PM, Chi JH, Dailey AT, Dhall SS, Harrop JS, Hoh DJ, Qureshi S, Raksin PB, Kaiser MG, O'Toole JE. Congress of Neurological Surgeons Systematic Review and Evidence-Based Guidelines on the Evaluation and Treatment of Patients With Thoracolumbar Spine Trauma: Timing of Surgical Intervention. Neurosurgery. 2019 Jan 1;84(1):E53-E55. doi: 10.1093/neuros/nyy362. PMID: 30202868.

59. Thanappan N, Saravanamuthu T, Singaravel, Sivakaran. Functional Analysis of Effectiveness of Short Segment with Index Vertebra Fixation as Compared with Long-segment

Fixation in the Management of Thoracolumbar Spinal Injuries. Int J Sci Stud 2019;6(12):146-150

60. Xu C, Bai X, Ruan D, Zhang C. Comparative Finite Element Analysis of Posterior Short Segment Fixation Constructs with or without Intermediate Screws in the Fractured Vertebrae for the Treatment of Type A Thoracolumbar Fracture. Research Square; 2022. DOI: 10.21203/rs.3.rs-2104845/v1.

61. Agrawal H, Sharma A, Singh V, Naseem A, Gaddikeri M, Amin A. Long-segment fixation versus short-segment fixation with instrumentation of index vertebra for thoracolumbar fractures. Surg Neurol Int. 2022 Jun 3;13:233. doi: 10.25259/SNI_238_2022. PMID: 35855118; PMCID: PMC9282815.

62. Jun li,Lei liu.Comparison of short vs long segment for the treatment of thoracolumbar burst fractures:a meta analysis.Int.J Clin Exp Med2017;10(2):1750-1762 ?????

63. Koh YG, Park KM, Lee JA, Nam JH, Lee HY, Kang KT. Total knee arthroplasty application of polyetheretherketone and carbon-fiber-reinforced polyetheretherketone: A review. Mater Sci Eng C Mater Biol Appl. 2019 Jul;100:70-81. doi: 10.1016/j.msec.2019.02.082. Epub 2019 Feb 22. PMID: 30948107.

64. Tezeren G, Kuru I. Posterior fixation of thoracolumbar burst fracture: short-segment pedicle fixation versus long-segment

instrumentation. J Spinal Disord Tech. 2005 Dec;18(6):485-8. doi: 10.1097/01.bsd.0000149874.61397.38. PMID: 16306834.

65. Dailey AT, Arnold PM, Anderson PA, Chi JH, Dhall SS, Eichholz KM, Harrop JS, Hoh DJ, Qureshi S, Rabb CH, Raksin PB, Kaiser MG, O'Toole JE. Congress of Neurological Surgeons Systematic Review and Evidence-Based Guidelines on the Evaluation and Treatment of Patients With Thoracolumbar Spine Trauma: Classification of Injury Neurosurgery. 2019 Jan 1;84(1):E24-E27. doi: 10.1093/neuros/nyy372. PMID: 30202904.

66. Sapkas G, Kateros K, Papadakis SA, Brilakis E, Macheras G, Katonis P. Treatment of unstable thoracolumbar burst fractures by indirect reduction and posterior stabilization: short-segment versus long-segment stabilization. Open Orthop J. 2010 Jan 15;4:7-13. doi: 10.2174/1874325001004010007. PMID: 20177428; PMCID: PMC2822149.

67. Basaran R, Efendioglu M, Kaksi M, Celik T, Mutlu İ, Ucar M. Finite Element Analysis of Short- Versus Long-Segment Posterior Fixation for Thoracolumbar Burst Fracture. World Neurosurg. 2019 Aug; 128:e1109-e1117. doi: 10.1016/j. wneu.2019.05.077. Epub 2019 May 17. PMID: 31103754.

68. Yung AW, Thng PL. Radiological outcome of short segment posterior stabilisation and fusion in thoracolumbar spine acute fracture. Ann Acad Med Singap. 2011 Mar;40(3):140-4. PMID: 21603733. ang JW, Hur H, Lee JK, Kim JH, Kim

SH.Predictive factors for a kyphosis recurrence following short-segment pedicle screw fixation including fractured vertebral body in unstable thoracolumbar burst fractures. J Korean Neurosurg Soc 2014;56:230-6.

69. Katonis PG, Kontakis GM, Loupasis GA, Aligizakis AC, Christoforakis JI, Velivassakis EG. Treatment of unstable thoracolumbar and lumbar spine injuries using Cotrel-Dubousset instrumentation. Spine (Phila Pa 1976). 1999 Nov 15;24(22):2352-7. doi: 10.1097/00007632-199911150-00011. PMID: 10586460.

70. Tezeren G, Kuru I. Posterior fixation of thoracolumbar burst fracture: short-segment pedicle fixation versus long-segment instrumentation. J Spinal Disord Tech. 2005 Dec;18(6):485-8. doi: 10.1097/01.bsd.0000149874.61397.38. PMID: 16306834.

71. Altay M, Ozkurt B, Aktekin CN, Ozturk AM, Dogan O, Tabak AY. Treatment of unstable thoracolumbar junction burst fractures with short- or long-segment posterior fixation in magerl type a fractures. Eur Spine J. 2007 Aug;16(8):1145-55. doi: 10.1007/s00586-007-0310-5. Epub 2007 Jan 25. PMID: 17252216; PMCID: PMC2200786.

72. McLain RF, Sparling E, Benson DR. Early failure of short-segment pedicle instrumentation for thoracolumbar fractures. A preliminary report. J Bone Joint Surg Am. 1993 Feb;75(2):162-7. doi: 10.2106/00004623-199302000-00002. PMID: 8423176.

73. Alanay A, Acaroglu E, Yazici M, Oznur A, Surat A. Short-segment pedicle instrumentation of thoracolumbar burst fractures: does transpedicular intracorporeal grafting prevent early failure? Spine (Phila Pa 1976). 2001 Jan 15;26(2):213-7. doi: 10.1097/00007632-200101150-00017. PMID: 11154543.

74. Canbek U, Karapınar L, Imerci A, Akgün U, Kumbaracı M, Incesu M. Posterior fixation of thoracolumbar burst fractures: is it possible to protect one segment in the lumbar region? Eur J Orthop Surg Traumatol. 2014 May;24(4):459-65. doi: 10.1007/s00590-013-1326-7. Epub 2013 Oct 5. PMID: 24091822; PMCID: PMC3990854.

75. BOUCHER HH. A method of spinal fusion. J Bone Joint Surg Br. 1959 May;41-B(2):248-59. doi: 10.1302/0301-620X.41B2.248. PMID: 13641310.

76. Roy-Camille R, Saillant G, Mazel C. Internal fixation of the lumbar spine with pedicle screw plating. Clin Orthop Relat Res. 1986 Feb;(203):7-17. PMID: 3955999.

77. Akbarnia BA, Crandall DG, Burkus K, Matthews T. Use of long rods and a short arthrodesis for burst fractures of the thoracolumbar spine. A long-term follow-up study. J Bone Joint Surg Am. 1994 Nov;76(11):1629-35. doi: 10.2106/00004623-199411000-00005. PMID: 7962022.

78. Knop C, Bastian L, Lange U, Oeser M, Zdichavsky M, Blauth M. Complications in surgical treatment of thoracolumbar injuries. Eur Spine J. 2002 Jun;11(3):214-26. doi:

10.1007/s00586-001-0382-6. Epub 2002 Feb 19. PMID: 12107790; PMCID: PMC3610522.

79. Li KC, Hsieh CH, Lee CY, Chen TH. Transpedicle body augmenter: a further step in treating burst fractures. Clin Orthop Relat Res. 2005 Jul;(436):119-25. PMID: 15995429.

80. de Peretti F, Hovorka I, Cambas PM, Nasr JM, Argenson C. Short device fixation and early mobilization for burst fractures of the thoracolumbar junction. Eur Spine J. 1996;5(2):112-20. doi: 10.1007/BF00298390. PMID: 8724191.

81. Dick W, Kluger P, Magerl F, Woersdörfer O, Zäch G. A new device for internal fixation of thoracolumbar and lumbar spine fractures: the 'fixateur interne'. Paraplegia. 1985 Aug;23(4):225-32. doi: 10.1038/sc.1985.38. PMID: 3900882.

82. Rommens PM, Weyns F, Van Calenbergh F, Goffin J, Broos PL. Mechanical performance of the Dick internal fixator: a clinical study of 75 patients. Eur Spine J. 1995;4(2):104-9. doi: 10.1007/BF00278921. PMID: 7600147.

83. Butt MF, Farooq M, Mir B, Dhar AS, Hussain A, Mumtaz M. Management of unstable thoracolumbar spinal injuries by posterior short segment spinal fixation. Int Orthop. 2007 Apr;31(2):259-64. doi: 10.1007/s00264-006-0161-4. Epub 2006 Jun 17. PMID: 16783550; PMCID: PMC2267576.

84. Gurr KR, McAfee PC. Cotrel-Dubousset instrumentation in adults. A preliminary report. Spine (Phila Pa 1976). 1988

May;13(5):510-20. doi: 10.1097/00007632-198805000-00014. PMID: 3187697.

85. Peters T, Chinthakunta SR, Hussain M, Khalil S. Pedicle Screw Configuration for Thoracolumbar Burst Fracture Treatment: Short versus Long Posterior Fixation Constructs with and without Anterior Column Augmentation. Asian Spine J. 2014 Feb;8(1):35-43. doi: 10.4184/asj.2014.8.1.35. Epub 2014 Feb 6. PMID: 24596603; PMCID: PMC3939367.

86. Lazaro BC, Deniz FE, Brasiliense LB, Reyes PM, Sawa AG, Theodore N, Sonntag VK, Crawford NR. Biomechanics of thoracic short versus long fixation after 3-column injury. J Neurosurg Spine. 2011 Feb;14(2):226-34. doi: 10.3171/2010.10.SPINE09785. Epub 2010 Dec 24. PMID: 21184641.

87. Reid DC, Hu R, Davis LA, Saboe LA. The nonoperative treatment of burst fractures of the thoracolumbar junction. J Trauma. 1988 Aug;28(8):1188-94. doi: 10.1097/00005373-198808000-00009. PMID: 3411642.

88. Xu GJ, Li ZJ, Ma JX, Zhang T, Fu X, Ma XL. Anterior versus posterior approach for treatment of thoracolumbar burst fractures: a meta-analysis. Eur Spine J. 2013 Oct;22(10):2176-83. doi: 10.1007/s00586-013-2987-y. Epub 2013 Sep 7. PMID: 24013718; PMCID: PMC3804690.

www.ingramcontent.com/pod-product-compliance
Lightning Source LLC
LaVergne TN
LVHW082354200926
848481LV00012B/3559

* 9 7 9 8 2 3 3 5 0 4 2 3 5 *